A K-12 Dance Method to Teach Creative Dance, English, Math and Science

Paul Moerman

A K-12 Dance Method to Teach Creative Dance, English, Math and Science

Letting Dance Teach

Springer

Paul Moerman
Arkeater
Rönninge, Sweden

ISBN 978-3-031-93155-0 ISBN 978-3-031-93156-7 (eBook)
https://doi.org/10.1007/978-3-031-93156-7

This Springer imprint is published by the registered company Springer Nature Switzerland AG
The registered company address is: Gewerbestrasse 11, 6330 Cham, Switzerland

To the children of this world,
and their teachers.

To Molly,
a dancer before a walker.

Preface

The strongest incentive to write this book comes from all the teachers, school leaders, and student teachers whom I have had the privilege to meet and dance with throughout the years, in schools and preschools, during seminars and workshops, at in-service training and conferences.

The teachers and student teachers have, on the one hand, asked for descriptions and hands-on lesson guides, in order to enact and further develop the dances in their everyday pedagogical activities. On the other hand, they have shown interest in the underlying artistic and educational ideas, as a rationale to advance the relevance of dance in their professional surroundings.

The book is a response to both needs. It inscribes itself in a proud tradition of providing dance method handbooks, through generations of devoted dance educators and scholars, teaching the young and doing research in academia. I deeply acknowledge the wisdom, know-how, and art of my forerunners. I particularly pay homage to one of the brightest torchbearers, Susan Stinson, whose indispensable work and writings have inspired my work, not least this book. In it, I seek to interweave Stinson's thoughts with John Dewey's and Gert Biesta's ideas, in an attempt to give their thinking on art and education body and motion and to further explore what dance can teach us about education.

Bringing an art mode like dance into the realm of school and education is no small matter. There is no point in inviting dance and, once there, deprive it of its transgressive force and its appeal to look at things from a different angle, for several reasons. Firstly, we would not be dealing with dance as the art form or subject matter it is. No math, language, or science teacher would imagine teaching their subject otherwise than to the hilt. They would not settle for some light version of their subject. Secondly, any curtailed variant of dance would not be fit to make any substantial contribution to the development of educational theory and practice. Thirdly, the very shift of perspective when working in an art mode such as dance is in itself a significant pedagogical element. A fourth point of bringing the arts into education is that the aesthetic comes with ethical implications, in life at school as in life at large, which is true not least in dance. This alone holds a big part of the potential for dance to contribute to what education is or should be: not a knowledge shop or a degree factory, not a marketplace but a meeting place where young people appear before each other and break into the world as subjects. This is where dance displays its true pedagogical force: children and students coming together and engaging in dance reveal education's relational and existential core in a tangible way. Dance is a mode of being in the world with others.

Hence, with this book I aspire to widen the gaze from current views on justifying dance and the arts in education, in terms of what they are 'good' for. All of them, however valuable, are utilitarian and therefore reductive of their power as modes of art. We are familiar with the arguments: dance as 'moving around' offers opportunities to get rid of surplus energy, for the benefit of better concentration in "real" subject classes; dance stimulates the brain for better learning; dance enhances fostering a fixed set of ethical values and better group dynamics; dance fosters creativity; dance offers a chance to express one's identity.

My ambition is to reach beyond these serviceable justifications of dance in education. The benefits listed are secondary and come into the bargain. Instead, I want to draw attention to the vital teaching force of dance in itself. I make a plea for a pedagogy taking the sensuous work seriously, a call for an aesthetic renaissance in teaching and education altogether. I hope this book will contribute to a deeper discussion on art and science, dance and education, aesthetics and ethics, the relational and the existential, education and democracy. Not until dance in school is given appropriate space to be dance undiminished: an art mode, a significant mode of knowledge and source of insight, an own meaning-making modality and a specific way of dialoguing with the world, will its force be able to operate fully in an education and a school for our time, and for our world's precarious future.

The book's discursive and describing text is interspersed with artist Christer Sundströms' pictures. His artworks dialogue with the text and hopefully with the reader. With their ephemeral technique—Buddha-board pictures are literally effacing—they breathe the sensuous and fragile quality of movement. Every move a young dancer makes fades in the same breath. Captured by the dancer's—and the beholder's—kinaesthetic sense, it lingers in the ongoing dialogue between the senses and the world of perceivable things and beings.

What happens in these fugitive moments and reverberates in the senses is what this book is about: young dancers breaking into the world as unique subjects in kinaesthetic interplay, attending, initiating, and responding to movement. What they carry along in their existence in the world in search of meaning and knowledge, what ethical choices they make for a fair and good life in and with the world, and whatever they choose to express in their dance, is beyond the pedagogue's quest.

Together, the text and artwork invite the reader to rationally and sensuously take in what the book tries to convey and may be encountered as aesthetic teaching. Ultimately, it is about what dance can teach us, from a broad educational perspective down to a singular move on the floor, in the flesh. When all is said and done, teaching and dancing with children, pupils and students is among the most wonderful, enlightening, overwhelming and joyous of gifts a teacher can be presented with.

So open your senses and let this book make a stand for dance *as* education: for what dance can teach—if we are ready to let dance teach.

Rönninge, Sweden

Paul Moerman

Competing Interests The author has no competing interests to declare that are relevant to the content of this manuscript.

Introduction

Education and the arts, like dance, are transgressive fields of action. Both deal with issues about the human condition, for the individual, for the citizen. The arts pose questions about what it is to be in the world. Education is fundamentally about efforts to enter into dialogue with the world. Interacting with our surroundings is how we become as subjects, as unique human beings, sensuously perceiving and participating in the social fabric of daily life in school, society and the global community.

Our world is constantly changing, at a dazzling pace. Never before has humanity faced greater challenges to take care of the world. Never before has the challenge been bigger to set about this task together. It appears to be our sole salvation. Our capability to maintain a sensitive dialogue with the social, natural, and material world is crucial in this quest. The central issue is not the individual person, nor even the human species, but what the world itself expects from us—for its survival.

It's us and the world. If not, neither might survive.

School is a world stage where we meet and interact with other people with different life experiences. When provided space to work in the arts, like dancing, young people can engage in ethically charged interplay with each other, without a given outcome. Difference and openness of perspectives are preconditions for democracy, where wills and minds are thrashed out without anyone claiming the upper hand. In an egalitarian society everyone is entitled space for initiative, for bringing their new beginnings to the world. When difference is seen as a resource, each and one can contribute to an ongoing, constantly changing historic and cultural process. This participatory perspective applies to the preschool child group, the classroom, the school, the nation, the global society.

To dance together is to carry on a dynamic dialogue of initiative and response in movement, filled with imagination and wonder. Dancing is attending to movement, one's own and others'. In numerous studies on dance in school, children, pupils and students emphatically word their dancing in relational and existential terms: an interplay marked by intense presence, heightened perception, joy and vulnerability, communication, closeness and empathy, trust and risk-taking. Their accounts are permeated with sensuous wonder, testifying to unexpected discoveries. Dance is tangible interplay and testimony to the interdependency between the 'I' and the 'Other'.

At the same time dance is a most realistic art mode, evolving in real space and real time. Its power of interaction comes out of the kinaesthetic experience of movement in all its abundancy of variation in space, time, body, and force. Moving and responding, the young dancers break into the world. Devoting to dance and immersing in movement—kinaesthetically experiencing—is a particular way of encountering the world. It is a specific, receptive work of the senses, preceding emotion and thought, paving the way for cognition, interpretation, creation, and communication of meaning and the encounter of specific knowledge, skills, and attitudes.

Teaching dance can be a model for aesthetic teaching in an aesthetic turn in education. Dance, putting the dialogical work of the senses central, has a potential to vitalise and challenge education in shaping a school for a future already here: a school with spaces open for (kin)aesthetic experiencing, filled with wonder, imagination, and empathy; spaces where young people can break into the world through their power of initiative and responsiveness, experience what it is to exist in the world, what gifts of insight and knowledge the world has to offer, what the world expects from us and how we, gaining insight, knowledge and judgment, can strive for a good life in and with the world, caring for our world together.

All this, dance can teach us, if we can imagine letting dance teach.

About the Book

Good education provides space for dance and the arts. 'Good' in the sense of a state-of-the-art education for the present and the future. Dance does matter to such an education, be it not primarily in terms of what dance and the arts commonly are said to be 'good' for, in a 'useful' sense. This book on teaching dance does not take a utilitarian stand, not on education and not on dance. It celebrates the delicate power, joy, and elucidation of dance in itself, and the educational relevance of dance in itself. It conceives a call for aesthetic teaching and an aesthetic turn in education at large. It is an ardent plea for an aesthetics of resistance to instrumentalising trends in education and the arts.

The book makes two prime points, both foregrounding the artistic and the aesthetic, and their intrinsic educational relevance for young people's lives.

Firstly, dance as an art mode is a proper mode of knowledge, insight, and meaning making. It must be taught as an aesthetic subject in its own right, with its own method—why the book presents hundreds of examples of dance activities. Aesthetics deal with what comes towards us from the world around us through our senses. The work of the senses is receptive and immediate, open to the unexpected coming one's way. It precedes emotion, cognition, interpretation, sense-making, and expression. In dance, a sixth sense is in force: *kin*aesthetics, sensing movement, in a dynamics of initiating and undergoing movement. The undetermined, the unpredictable, and even the unknowable are part of all art making and art experiencing and must be part of teaching an aesthetic subject like dance. It calls for an aesthetic teaching mode challenging control, linearity, and measurability. In educative terms, the first question the book deals with is all the wonder the young person can 'be taught' by engaging in dance and kinaesthetics and embracing the unforeseeable.

The second point made, broadening the perspective, is that coming together and dancing in school can remind us—'teach' us—what education is about: young people's efforts to find a place in the world, come to terms with the world, enter into dialogue with the world—our wonderful, challenged and challenging social and natural world. Artists and dancers ask questions about our human condition, trying to figure out what it is to be in this world. Taken into the realm of education, dance as an art form must maintain this ultimately existential perspective: me and the world, which indeed is also the fundamental question of education. This second question informs the book's theoretical introduction, bringing contemporary educational theory and dance educational theory and practice together.

While the former point spotlights dance as a singular aesthetic subject, the latter highlights an existential aspect common to dance and education. Together, they form a firm aesthetic, ethical, and existential argument why dance is essential to a qualitative education for our and future times. Dance must be offered space to operate in education with all its transformative force, vulnerability, and joy. Dance can enrich and reinforce what education is all about: young people entering the scene, gaining knowledge, proficiencies, attitudes and insights, growing, changing, finding their own meaning. All through, good education must provide a platform for young people's becoming as subjects of their own lives, ultimately making ethical decisions in challenging issues, for themselves, in dialogue with the social and natural world they inherit and will hand over to new generations.

All in all, this book is an invitation to look at teaching and education through the dancer's eyes and to look at dance through the educator's gaze. The book invites the reader to look at dance *as* teaching and to be willing to let dance teach.

Between the Covers

The book consists of three interconnecting parts: a theoretical introduction, a method handbook, and a concluding summary.

The introductory Chap. 1 presents a theoretical foundation for the teaching of dance. Attention is drawn to the intertwinement of the artistic and the educational, the aesthetic, the ethical, and the existential. An ontological and epistemological framework is formulated, mainly drawing from educational theorist Gert Biesta, educational philosopher John Dewey, and dance scholar and educator Susan Stinson.

The method section is the lion's share of the book. Chapter 2 deals with the teaching of dance, exploring the elements of dance, improvisation, and composition. Chapter 3 elaborates on the teaching of creative dance, guiding further into the processes of improvisation and composition, and the creation of choreographies. Chapters 4, 5, and 6 deal with teaching dance as an aesthetic teaching method to teach English as first or second language, mathematics, and natural sciences.

Each chapter in the method section opens with a short abstract linking to the theoretical framework. Lesson plans are presented, with numerous dance exercises relating to single learning objects or themes. Each dance activity is followed by suggestions to structure a *meta-talk*, where the subject content dealt with is defined and named in appropriate language. With the abundancy of examples as support and inspiration, the reader/teacher is encouraged to develop own ideas together with their young dancers, and expand their dance teaching.

The final Chap. 7 summarises the book's teacherly mission around the initial question what dance can 'teach'. Drawing upon the dance teacher's grateful task to initiate young people into the magical world of dance, 'gifts' of teaching are ascertained. The concept of an aesthetic education is outlined: an education taking the work of the senses seriously, along with its existential task. An education preserving young people's integrity, providing space to become as subjects of their own lives, acting upon their freedom of initiative, receptiveness and imagination, in dialogue with the social and natural world. A world ready to offer its 'gifts' of knowledge and insight to the sensuously receiving 'I'.

Teaching, in conclusion, is forwarded as apt to allow the world to reveal itself aesthetically—a vision Herbart, the first pedagogue, envisioned as the *art* of teaching already three centuries ago.

Target Groups

With its combined theory and method content, the book is directed firstly at student teachers and teachers in preschool, elementary school, inclusive education and before-and-after-school extracurricular activity, in search of ways to expand their teaching of dance, and dance integrated in the teaching of language, math and science. Secondly, it appeals to everyone concerned with education and the arts, from politicians and decision makers on national and local levels, to principals, peer teachers, and parents.

All in all, the book hopefully speaks to all those interested in the deeper meaning of dance and the arts for pedagogy and for young people's education, learning, development and becoming as subjects in and with the social and natural world.

Contents

About the Author

Paul Moerman is a dancer, an actor, and a literary translator, educated at Stockholm University of the Arts and Valand Academy at the University of Gothenburg, Sweden. He is a teacher educator at Södertörn University, Stockholm, Sweden and a doctoral candidate in Arts Education at the University of Jyväskylä, Finland. Formerly a civil engineer and an architect educated at Ghent University, Belgium and the Royal Institute of Technology, Stockholm, Sweden, he has worked as an architect in Belgium, Sweden, and the USA.

Along with his artistic work on stage and film, he has pursued a teaching career in dance, drama and creative writing, as artistic expressions in their own right, and integrated these in the teaching of language, math, science, and philosophy. He has been teaching for 30 years in preschool, elementary and secondary school, inclusive education and after-school activities, and at teacher educations at universities in Sweden, Finland, the Netherlands, and Australia.

He has designed a teaching and learning methodology drawing on dance, featuring the programs *Dancing through the Alphabet, Dancing Math, Dancing Natural Sciences* and *Dancing with each Other.* He has published and presented his research on dance in/*as* education worldwide. Modelled on his practice of dance teaching/*letting dance teach*, he formulates a concept of aesthetic teaching, advocating an aesthetic turn in education.

Part I
Teaching Dance: A Theory

Teaching, from early childhood education, is a professional discipline informed by pedagogical theory and everyday practice. Its foundation is like a soil, to use a metaphor evoked by the German loanword Kindergarten, in need of nutrition with fresh teacherly practices and new ways of looking at the pedagogical work. What we teachers do in refining our daily work with young people is to let in air and give space for them to thrive—as co-gardeners in sensuous resonance with their natural and human surroundings. How we conceive and view our work contributes to the maturation and further development of the educational discipline's theory and practice.

This goes not least for the work in the art modes in schools, and certainly for dance. In many national curricula and teacher education syllabi the arts are formulated in general terms of global educative goals, and often treated collectively. A lot is yet to be done to define the nature and relevance of dance in education more substantially.

The ambition in Part I is to contribute to a language and a theory of knowledge to more elaborately discuss and better understand the what, why and how of teaching dance in education for young persons' becoming, change and growth. Underlying the following argumentations is a conviction that dance teaching must be directed at creating space in schools for the young dancers to thrive—as choreographers.

Chapter 1
Ontology and Epistemology

P. Moerman, *A K-12 Dance Method to Teach Creative Dance, English, Math and Science*, https://doi.org/10.1007/978-3-031-93156-7_1

Dance, a thoroughly tangible, aesthetic activity, is infinitely grateful to teach. A child can fill the tiniest movement with presence, wonder and magic. What they enact in these single moments—and is a privilege to witness—is what dance in education is all about.

In formulating a set of concepts and a theory of knowledge to frame an academic discussion about dance teaching, the entwinement between the aesthetic and the ethical is central. This in turn is tied to the relational and existential aspects of both dance and education—dance understood as an interplay in movement, and education as trying to enter into a dialogue with the world.

Dancing in school, and working in the arts in school in general, in all their tangibly aesthetic qualities, can remind us of what education is or should be about. This is especially valid in times of control, rationalisation and marketisation of education, stressing empirical knowledge transfer, and competition—between nations, schools and students. In this, a hierarchy of subjects is set up in which dance and the arts are rarely high-ranked. In our same era, young people urgently call for joint efforts and new perspectives, indeed for radically new ways of living, to solve pressing local, national and global problems, society and nature related. Their call is of great ethical and existential dignity, for the survival of democracy and the planet.

It may sound overly grandiose to initiate an ontological and epistemological consideration of dance and the arts in schools by referring to existentially charged social and material challenges in society and the world. The question is, again, if this is not precisely the whole idea of teaching and working in the arts in education. Asserting them as distinctive sources of knowledge and insight, the question is if dance and the arts are not indispensable in an education for the present and the future.

Schools are in the middle of our stirring era. Yet, many children and young people around the world feel a weak sense of belonging at school, because of their living conditions and lived experiences—many are deprived of school altogether. By contrast, many other young people, who do feel included and do realise school's value and significance for their lives, nonetheless take the drastic decision to go on school strike, directing our attention to those truly burning life issues. What they do is what artists do: to make us stop and consider our doings, and confront ourselves with the larger picture of what it is to be in the world.

In essence this is also what the educator does when teaching: to point out something in the surrounding world worth paying attention to, of importance to our life in and with the world. By pointing, the pedagogue summons the young individual to turn to the world, step forward as subject, dialogue with the world of things and beings and add their response to issues of interest to us all. Working in the art modes offers a form for such a dialogue with the world, touching upon the existential—what it is to be in the world.

Creating spaces in school to work in the arts is creating spaces to hold on and figure out what to do with all the knowledge, skills and social attitudes we gain in education. Spaces to meet even the unknown, the not beforehand determined and the possibly unknowable. Spaces for freedom of initiative and freedom of thought. Spaces of becoming as subjects capable of acting on that freedom, for the good of one's life and the world.

Below, these threads of thought are woven together in an exploration of what dance in its particular and entrancing way can teach us about education as trying to be in and with the world. Subsequently, the discussion suggests how dance, when operating in education, can contribute to the development of educational theory and practice, in terms of general teaching and learning as well as subject teaching and learning. The discussion also has implications for a qualitative teacher education and a clearer statement in the syllabi of the nature and relevance of working in an art mode like dance, in its own right as well as in teaching other subjects.

The emphasis in this account is on building aesthetic and ethical relationships in dance and education, and on entering into dialogue with the social and natural world, giving the young person opportunity to discern, amongst all their aspirations, what the world may wish from each one and what each one can bring to the world.

The dialogue evolving in dance and the arts is described in terms of aesthetic experiencing, a passage of becoming as subject, and a pathway towards insight and knowledge. In an aesthetic experience the work of the senses is central. It is a mode of experiencing marked by sensuousness, wonder and surprise. It is also a mode of ethical interaction in which plurality and intersubjectivity come before identity and expression.

In dance the dialogue is *kinaesthetic*, unfolding through kinaesthetic experience: to sense movement, to take initiative and respond to movement. To dance is to open one's senses to the world, letting the world reveal itself aesthetically and offer its gifts to the one willing to receive its gifts. The work of the senses precedes cognition, emotion and expression. Ultimately, dancing together in school is seen as encountering a world of other and others, and therethrough encountering oneself.

Dance is at the same time the most magic and the most realistic of art modes, each step coming into being in the flesh, filled with meaning of and in itself. From the first step, exploring dance's elements space, time, body and force, the young dancer immerses in sensuous exploration of the world. This chapter inquires into what attending to movement entails in a broader educational, relational and existential perspective.

Dialoguing with the World

Educational theorist Gert Biesta (2017) observes a number of striking similarities between the arts and education as fields of activity. While the arts deal with queries about what it is to be alive in the world, education fundamentally deals with young people's efforts to enter into a dialogue with the world. To maintain a dialogue with the world, says Biesta, is to exist as subject in a grown-up way: not aspiring to be in the centre of the world, but to try to be in the world along with others and the otherwise.

In artist Joseph Beuys' iconic happening in 1965, *How to explain pictures to a dead hare*, Biesta sees the staging of the archetypical form of teaching: to *point* at something in the world for someone, something there to perceive, worth paying attention to, meaningful for one's life in the world. The teacher's pointing gesture is pedagogical, and as such is thoroughly relational and situated: it goes from this very teacher to this very pupil and that very thing in the world which might be significant. Furthermore, says Biesta, Beuys performs the archetypical modality of teaching: to *explain*: an invitation to the pupil to be initiated into the realm of meaning making.

Moreover, Biesta remarks, in his performance Beuys explains *pictures*, the picture being the archetypical object of teaching: something visible, perceivable by the senses. Biesta arches back in the history of pedagogy to Johann Amos Comenius (1991/1658), who had the ceremonial ambition to present all sensuously perceivable things in the world in his *Orbis sensualium pictus*, the first textbook including pictures as means of explanation. Comenius wanted every child to be able to go to school, poor or rich, girl or boy.

Through the archetypical gesture of teaching, to point, and its archetypical modality, to explain, the teacher offers the pupil the keys for searching for truth and meaning. This is a gift without counterclaim, Biesta argues, lest the teacher wants to control the process of teaching 'at the other end'. To respect the pupil's freedom is to leave it to them to step forward as subject, to pay attention, willing to be taught—or to stay back as object.

In summary, the teacherly gesture, akin to the artist's, consists of pointing at something worthwhile to someone, turning the young person towards the world: 'Look, there!' It is an effort to awake a longing in the pupil to be in the world in a grown-up way: not putting themselves in the centre of the world, asking the world to fulfil their desires, but to bring themselves to ask the world what it may wish from them. It shows a pathway to dialogue with the world.

It is an arduous and probing path to wander, Biesta argues, through disturbance and resistance. It calls for a discernment which of one's desires towards the world might be more or less desirable from the world's point of view, a world unable to accommodate all our desires: a world where I am not alone. In turn, it is a more rewarding path in the long run, leading towards new experience, insight, knowledge, enlightenment, change and judgement.

Working in the art modes can show us such routes, Biesta suggests. Dancing, working with material and abstract elements like body, space, time, force and relationships, dance educator and scholar Susan Stinson (1988, 2016) states, can show the young dancer that these things not only exist in dance, they also exist in their surroundings and the world. They can shape fluffy forms like clouds do, they can move swiftly like the wind, they can move like a classmate and move on together. While dancing, they learn to know themselves in interplay with the world. Dance can 'teach' us how to enter into dialogue with the social, natural and material world.

Becoming

When elucidating education's relational and existential dimensions, a core issue concerns what space is provided in education for the individual to emerge as a unique subject.

Biesta (2006, 2021) advocates an education not solely concerned with transferring predefined knowledge, skills and values from one generation to the next, following fixed curricula. Rather than an education taking on the task to foster and mould the young individual towards fulfilling universal ideas of humanism—commonly the Kantian ideal of the autonomous, rationally acting individual –, Biesta argues for an education that cares for everyone's uniqueness as a human being: a space for difference. It is a space where everyone has the equal right to participate and to be listened to about issues of common interest.

We reveal our singular subjectivity, Biesta says, in the way we respond to others, more so, in how we take responsibility to let others emerge in all their otherness. Biesta leans on to philosopher of ethics Emmanuel Levinas (1992), who claims we come into being as subjects by assuming ethical responsibility for the *Other*, the one I am not: the ethical precedes existence, we are *for the other* before we are *by ourselves*—a self. In response, school must be a true place for difference in order to be a place for each and every one to become as human subject.

Diversity is a precondition also for democracy to come into being. Biesta refers to philosopher Hannah Arendt's (1954) thinking on freedom, which she ties to the human capacity to take action, to initiate—to make an *initium*: a new beginning. Humans disclose their uniqueness in words and deeds by bringing their beginnings to the world, adding something that was not there before. What is initiated is however not realised until it is received by others. Becoming as *subject* unfolds alternating between being the *subject* of one's actions and *subjecting* to others' response to one's actions—others equally capable of taking initiative. And so, an intersubjective space is established, ruled by freedom of action and conditioned by plurality.

Freedom, Arendt underscores, occurs *in* the very act of bringing about something new, and presupposes the presence of others. She refers to the performing arts, where the artwork is accomplished in the performance itself, in the very acting, not in a tangible outlasting product. The actors or dancers on stage depend on each other's reactions here and now, and on an audience to witness these momentary acts. Freedom, Arendt stresses, can only occur in the intercourse with others, where no one has the best of it.

Hence, concludes Biesta, plurality and diversity shift from posing a problem to be solved, not least in education, to an asset, a resource in creating a community of scrutiny and difference, a space of becoming and being-with-others.

Questioning an education with a one-sided emphasis on learning as acquisition of fixed knowledge and values, Biesta (2006, 2021), inspired by Levinas (1969) posits a take on 'being taught' by being addressed by the unknown, the unexpected and the unpredictable, and responding to such encounters. The others and the otherwise offer moments of interruption, disturbance, and a loss of sovereignty. An educational opportunity occurs, calling for risk-taking and to linger and labour in the not-known. What education first and foremost must cater for is the young person's subjectivity, and create contexts where they can try their way ahead, break into the world and take place in the social weave of difference with their own thoughts and feelings.

The educational question then, concludes Biesta, is not 'Where am I?' but 'Where are you?' That question calls *me*, as a unique being, as *someone*, to the world. It widens education and learning from being understood as learning something existing, to learning by responding to the unknown and the uncertain, that what is open-ended, troubling and challenging. Learning turns into *creating*, inventing. Education turns into a space of becoming as unique subjects, a space for freedom of action and the right to participate, to listen and be listened to in issues of common interest, bringing one's unique answers to the world.

Dancing as Dialoguing and Becoming

Dance is an art mode, and as such is a distinct way of being in the world. It is literally a way of *moving* through the world, as Stinson (1990) expresses. Dancing always evolves in dialogue with the world outside the studio, the stage or the school—preschool and college alike. Dancers relate, not seldom in opposition, to lingering ideas about art and dance, about who can dance, who is allowed to dance, about body and gender, social communities, knowledge and power. These issues follow along when dance is taken into education. Ethical aspects come with the aesthetic, when young people dance with and in front of each other, adapting or resisting to views on dance.

Contemporary dancers interact with the world through dance as an *aesthetics of resistance* (Manchev, 2011) to all kinds of objectifications of the body: as commercialised item, as lifestyle fetish, as projection surface for aesthetic, social, political, gender and even biological codes, or as a movable puppet in the hands of a genius choreographer. Instead, present-day dancers interact as subject-bodies in sensuous communication without a given outcome, like living windows wide open to the world, exploring experiences 'not yet one's own'.

This is precisely what children do dancing, says Stinson (1988, 1990): exploring the magic in movement, they break into the world by force of imagination, inventive bodies eternally becoming as subjects. Hence the dance teacher's task, Stinson marks, is to create space for such discovery and exploration, without given ends, for freedom of trial, for risk-taking, a process where all involved learn about themselves, each other and the world we live in. Dancer and philosopher Erin Manning (2009) views every movement in itself as a new beginning through which the very body becomes, dancing 'the not yet'. Every movement is body and dance coming into being: dancers dancing the dance of becoming.

The images of the child dancing their way into the world and the dancer coming into being by moving along with a movement capture the essence of dance as existential, relational, body-sensuous interaction preceding emotion, cognition and consciousness, holding a meaning in itself. The dancer emerges as subject initiating and responding to movement, creating an unpredictable, boundless space of appearances and beginnings—an irresistible intersubjective space.

Aesthetic Experience

Interaction, participation and communication are the actions constituting an *experience*, the central concept in pragmatic philosopher and progressive educationalist John Dewey's (1916) thinking. An *experience* is the result of the transactions with the surrounding physical and social world through which the individual gathers knowledge and insight into some aspect of being in the world: *learning by experience*. Experiencing is an ongoing process of *doing* and *undergoing*, a reciprocation between action and submission to the consequences of one's actions. Thinking and knowing progressively come out of unceasing interaction with the world.

In *Art as Experience*, Dewey (1934) elaborates on the experience concept in aesthetic terms: what one undergoes and enjoys in the presence of a work of art—architecture, poetry, music or stage art. The beholder engages with the work, seeks to grasp its form, content and meaning. Through intense perception they perform a work similar to the artist's, a sort of re-creation, which Dewey (1938) also calls *recreation*, suggesting to view art as a form of play.

Through the work of the senses, the aesthetic penetrates the experience from without, from the art work. The work is laborious, the beholder takes initiative, gets frustrated, does not get a grip of it at first attempt, loses a foothold, goes through phases of suffering, disturbance and dissolution. The aesthetic comes out of the clarifying and intensifying treats of the consummation of the experience, as a whole is created from what at first appeared as separate fragments. Meaning is created and communicated between the beholder and the object of the experience—the art work. An understanding emerges, the work of art is literally rendered significance as the physically divided parts merge into a whole. Resistance, tension and emotion turn into appreciation, satisfaction, participation, insight and growth as the experience is consummated.

While the encounter with artwork models aesthetic experiencing, Dewey (1938) points out it can as well occur in the presence of nature, everyday objects and scenes, with people, in mathematics and science. If we are sensitive and receptive to such experiences, they can strike us with a genuinely aesthetic quality, and enlarge us. All experiences have, at best, an aesthetic quality, says Dewey, why we could strike the adjective 'aesthetic'.

When aesthetic, an experience stretches beyond the emotional and intellectual, it is no symbol or representation nor has it a significance beyond itself—the only way to have the experience is by *having* it in its vital totality. The interaction with the object of the aesthetic experience, the labour of attending to it, constitutes the very experience. It is situated in space and time, in a here and now, and simultaneously incorporated in a chronology of previous experiences. It is corporeal and there is always movement involved, it appeals to our senses: our eyes, ears, sense of corporeal movement, touch.

Only through sensuous perception can we have a direct, original communication with the world, underlines Dewey (1938). The work of the senses evolves in a union of passivity and activity, of receptivity and responsiveness, a sensitivity to the world of persons, things and natural events that is lost when controlled by reason and cognition. If the intellect looks at them as signs of something beyond themselves instead of what they are, they cease to be aesthetic, they cease to be the works of art our senses, doing and undergoing, make them into.

Aesthetic Education and Aesthetic Literacy

Expanding on Dewey's concept of aesthetic experience, educational philosopher Maxine Greene (1986, 1995, 2001, 2007) develops her ideas on aesthetic education and aesthetic literacy. Aesthetic education is concrete action, fostering through intensive, receptive and reflective encounters with the arts—which are always new beginnings. Children and students engage in them with all their senses, connecting to them with a diversity of meaning. Aesthetic literacy is a way of thinking, probing unfamiliar tracks of thought and new ways of structuring lived experience. The pupil sees more, hears more, feels more and senses more, makes unexpected connections and investigates deeper layers of meaning.

Greene (1986) compares spaces where aesthetic education can evolve with Arendt's notion of the public space, presupposing the presence of others, appearing in front of each other with the best of their capabilities and in all their diversity. Only active, growing human subjects can establish such spaces, coming together, initiate new beginnings and achieve their unique subjectivity.

Greene (1995) views imagination as a social capacity which helps us get a grip of a complex world, enabling empathy, helping us to cross the empty spaces between ourselves and 'the others'. Imagination allows us to make alternative realities credible, breaking off what we take for granted and laying ingrained distinctions aside. Aesthetic education enhances a reflective response and a capability to experience such a response from another person, a response Greene denotes as ethical, opening for change.

Dewey models aesthetic experience on encounters with works of art, architecture or poetry. Greene advocates an engagement in the raw material of the arts: canvas and colours, tones and parts, poems, bodies in motion. Moving with dancers, balancing one's weight or feeling the effort-shape of an outflung arm will help perceive more profoundly what unfolds in a dance performance. Neither deepen into aesthetic experiencing in the young person's own creative work in the arts.

In his argument about what the arts can teach us, Biesta (2017) delves all the more into the student's own doing of art work: the work of the hand, the head and the heart encountering the realities of paint, stone, sound, body—once called 'resistant materials' in the British curriculum. Resistance—the point at which education begins—causes disturbance and interruption in the student's existence, which offers golden opportunities to scrutinise and sort out which of one's desires might be more desirable in meeting the world. The teacher's task is to 'stage' the encounter of the student with the world, and to prompt them to linger in the difficult 'middle ground' between going about too hard and shying away, between destroying the world we seek to exist in and self-destruction, obliterating our very existence in the world. Enduring and coping with the effort it takes to exist in and with the world, holding back the urge to be in its centre, is trying to be in dialogue with the world and to be there *with others*: a world-oriented take on education rather than a self-oriented (Biesta, 2021).

The middle ground is a thoroughly worldly and educational space, Biesta (2017) underscores, it makes us meet the realities of the world as well as the reality and the limitations of our own existence in the world. It teaches us we are not alone. It offers space of reconciliation with reality. The work in the arts offers space for play and trial of limits to our ideas and aspirations, to find out what it is to be in dialogue with the world and to find aesthetic forms for this dialogue. Visual art, music, writing, drama and dance are each and one specifically distinct, sensuous channels to enter into dialogue with the world.

Kinaesthetic Experience

The dance teacher's rewarding mission is to escort the student into the wonderful world of dance. Through a rich repertoire of play and exercises, exploring movement quality and movement variation, the young dancer is brought to attend to movement: to become *kinaesthetically aware*. The kinaesthetic sense, which Stinson (1988) calls a sixth sense, is what turns movement into dance. *Kinaesthesia* comes from Greek *kinéin*, to set in motion, and *aisthesis*, sensuous perception. It emphasises a union of power of action, *to initiate* movement, and a sensitive capacity to *sense* and *experience* movement. It adds a *kinetic*—a movement dynamic—dimension to Dewey's aesthetic experience concept, a corporeal and sensuous aspect of doing and undergoing, and an emphatically relational one.

For a long time, Stinson pondered how to capture in words the connection between the aesthetic and ethical dimensions of human existence with reference to dance education. After a walk taking a break from writing, she laid down to rest and arrived at a theoretical framework for her research: it is about the difference between the vertical and horizontal position of the body. The upright position stands for activity, control, mastery and command—to be *on top*—while the extended position stands for passivity, susceptibility and docility—to be *with* others.

We may also view the vertical posture, like in the walking brake, as reposed and beneficial to the free flow of thought. New insights and angles take seed while we stand still, walk, lie down or shift positions. Alternating between the 'upright' and 'extended' body is a fitting metaphor for the dynamic between the dancer's giving and receiving of movement impulses, for kinaesthetic experience in dancing with others.

Biesta (2017) borrows the notion *passibility* from applied cognitive scientist Wolff-Michael Roth (2011), who pairs it with *affectedness* and *passion*. It refers to the capacity to be affected through the senses, central in daily experience, and making learning and understanding possible. Sensuously affected in encounters with the world, Roth (2011) claims, the living/lived body is aware of being able to be further affected, without mediation of a conscious mind, and, willing to move on and be further affected, changes and develops. Roth defines passibility as a blend of agency and radical passivity, a willingness to be permeated through the senses by the not-yet-known, the result of active, non-cognitive decisions—in turn enabling cognition.

This touches upon the radical potential and relevance of dance in education, what dance can teach us, and how dance may inform aesthetic teaching and an aesthetic turn in education. Kinaesthetic experience presupposes the capacity to experience and surrender to movement: to follow along with the fling of an arm and let it take one further, let the arm, the shoulder, the torso, the hip be caught by the movement, find a new twist in some direction and move along again, on and on, before thought enters and steers the next step and gesture. What is dealt with here, in educational terms, are modes of knowing and understanding, capacities, skills and attitudes different from what is at stake in teaching and learning predefined subject knowledge and concepts. What evolves here is pre-cognitive receptiveness to and action upon the senses, including the kinaesthetic.

Kinaesthetic experiencing, says Stinson (1988), is the sensuous internalisation of movement through which we can sense others' dance, and our response to it. We become participants, we breathe with those we see dancing, we feel the stretch continuing from the tips of the fingers, we feel the body gently landing after a leap. Kinaesthetic experience constitutes both the active and the receptive part in the work in the art mode of dance, which, in Biesta's (2017) terms, is a specific channel for dialoguing with the world.

Teaching Dance as Aesthetic Teaching

The capacity to be moved by movement is the kinaesthetic translation onto the domain of dance of Roth's concept *passibility*, the capacity to be affected by what comes towards us through the senses: what 'gives' itself to our hearing, our seeing, our sense of touch. The capacity's bond of radical passivity and potentiality to act upon the 'given' is akin to Stinson's view of the horizontal body position as an image for being and moving with others, and how the aesthetic and the ethical relate in dance teaching. It is likewise allied with Dewey's view on the work of the senses in aesthetic experiencing, a union of passivity and activity.

These views resonate with Levinas' (2003/1972) stands on vulnerability as a radical passivity, more passive than all passivity, located entirely in the body: 'a nakedness more naked than the naked of skin'. Like Roth, Levinas explores the self's encounters with the world and what comes towards us in bodily terms, literally in the flesh: the body is a *sensing sensed*, combining perception and expression, creating language, poetry and dance, giving meaning to what is 'given'. To perceive is to receive and express at the same time.

To lend an ear, open an eye is to attend to what is 'given' to me: to respond, a first: 'Here I am!' In kinaesthetic terms: while dancing, I let myself be 'summoned' as an 'I' by the movement coming my way. My receptivity, undergoing the movement, is my first response as *subject-body*. In ethical terms, facing[1] another human being, asserting and assuming responsibility for the other's vulnerability, I respond: 'There you are!' The receptivity at stake is a sensitivity preceding interpretation, will, thought and emotion. These are acts of becoming: confronted with my subjectness in front of beings and things coming my way, I make a choice—to be an 'I'. It encompasses a risk-taking to be struck by the other and what is other, by the unfamiliar and the unknown. Levinas (1969) calls it a willingness 'to be taught': to receive more than I can contain within the totality of my perception of the world—a sense of infinity.

[1] Levinas (1987/1992) pictures the encounter with the other as a *face* emerging, not metaphorically, but in the flesh: a defenseless face approaching me. The encounter occurs in proximity, through listening and speaking, body language, facial expression, glances.

The readiness 'to be taught' what is beyond my capacity and control is opposite to a teaching order in which both teacher and 'learner' are in control of 'learning' processes and outcomes, sticking to predetermined knowledge and keeping away from the risky and the unexpected. Control so infringes on the freedom of the student's 'I' as it objectifies the student in the educative process. It also obliterates the student's chances to surprise the teacher, and themselves. It is an order putting a premium on cognition before sensuous perception—of 'thought control in the classroom'.[2]

These strands inspire a concept of *aesthetic teaching* and a call for an *aesthetic turn* in education: an education taking the work of the senses seriously, taking the work in the art modes seriously, and taking education's existential task seriously.

Dance, all said and done, is for real. Dance evolves in the real world of bodies, in the real world of others. As modern dancer and choreographer Merce Cunningham (1955) remarks: flows of dance movements evolving in time and space are fully visible to anybody—dance is the most realistic of things that occur on a stage. Yet creating magic. Yet yielding real new beginnings, which every dance teacher witnesses time after time in awe.

Crucial in creating space in school for dance and the arts is to allow space for the unfamiliar and the otherwise, for the unpredictable, for new, unexpected perspectives of being in the world, for what escapes cogitation—ultimately for what the world may offer as 'gifts'. Reasoning on the 'gifts of teaching', Biesta (2021) refers to philosopher Jean-Luc Marion, who takes a radical leap in the phenomenology of human experience: he situates perceivable things before the 'I', who, 'seeing without foreseeing', open to the unpredictable and the unknown, 'receives' what comes one's way.

Education keeping away from the unexpected and the unpredicted jeopardises the freedom of the young person's 'I', facing what possibly may come forward from the world as a richness of 'gifts'. To see without foreseeing is more moving than just 'seeing' things, it is about letting oneself be called to come experience what may be there, ready to give itself to me. It is a subjectification both of myself experiencing and of that what offers itself by its own initiative (Marion 2017). It is different from knowledge acquisition as going out to the world as a storage and go grab hold of what I desire.

This order, says Biesta (2021), turns school into a place of revelation, a realm beyond one's horizon of understanding, with a potential to strike with new, unexpected insight. More than we wish to learn about the world, the world is trying to give itself to us. Receptive, ready to be given, and receiving through our senses, we let ourselves be summoned to the world. Dancing, doing and undergoing movement is a model for this worldly dialogue.

Pointing as the archetypical gesture of teaching, Biesta (2017, 2021) proclaims, goes from the teacher to the student and the aspect of the world worth paying attention to. The figure also describes *the didactic triangle*, coined by Herbart (2017/1902). The pointing gesture aims at turning the student towards the world. The explaining

---

[2] Like the child choir sang with Pink Floyd on the album The Wall 1979, in a call for a ban on caning. The question asked here is if there still is a need to tear down walls around schools to let in air and preserve young people's true integrity, subjectivity and freedom of initiative.

act offers the keys to recognise the truth as truth. Together they invite the student to be initiated into the realm of meaning making. It is the teacher's task to make the invitation tempting. They can apply a range of didactic means on the one hand, following a common order of controlled instruction aimed at intended outcomes.

But the dialogical work can also be 'subjected' to the work of the senses, by rendering the teacherly work an essentially aesthetic dimension. Biesta (2024) claims, with Herbart, that to get a response to the invitation from the student, and for it to be truly the student's own and free response, the (re)presentation of the world needs to be aesthetic. Herbart (2017/1902) sees this as the *art* of teaching: to allow the world to reveal itself aesthetically.

Dance is a (kin)aesthetic way of being and moving through the world. The dialogue as described in this artistic and educational discourse, at one and the same time pedagogical, aesthetic, ethical and existential, is one between the young dancer ready to be initiated into the world of dance and movement, a *sensed* subject-body *sensing*, perceiving and expressing—creating dance—receiving and giving meaning to the gifts offered by the world.

The envisioned aesthetic teaching and the aesthetic turn in education evolve in terms of the teacher trying to awaken a desire in the young person to turn to the world and to be there as subject in sensing dialogue with the world-as-subject, open to 'subject' to the world's gifts. The receptiveness at stake is neither passive nor primarily cognitive. As dance teachers know, the work is laborious. It goes, as Biesta (2017, 2021) says, through resistance, interruption and delay, even passion and suffering, and the young dancer may need support to linger in this difficult, vulnerable middle-ground between the self and the world, between self-destruction and world-destruction.

Working in the art modes shows us such a path to dialogue with the world, if we are ready to let art 'teach' us that. Through the work of the head, the hand and the heart we meet the world, and meet ourselves in the world. Yet facing limitations of bodily movement, yet surrendering to the flow of movement, without a given outcome, caters for such encounters, and opportunities to letting dance teach.

Letting Dance Teach Language, Math and Science

In teaching creative dance the aim is to cater for open-ended exploration of movement, for the young dancer to become kinaesthetically aware, to sense movement and find movement's own significance, and with new insights and skills create choreographies.

Kinaesthetic experiencing is also 'the motor' in the teaching of dance in teaching language, math or science as conceived in this book. Dance instruction is the starting point. Many dance exercises presented in those chapters are borrowed from the teaching method of dance in its own right, while other activities are specially designed. The 'other' subject's content is dispersed in the dance activities, clearly perceivable or more concealed. The interaction in movement on the floor unfolds in multiple, interconnected layers.

Firstly, the students are offered opportunities to become kinaesthetically aware. Secondly, their kinaesthetic literacy, to 'think' like a dancer—a 'thinking' this book attempts to broaden beyond the realm of reason and cognition—is brought to interplay with their linguistic literacy, mathematical literacy/numeracy, or scientific literacy, to think like a linguist or a writer, a mathematician or a scientist—their 'thinking' made possible by the capacity to be sensuously affected. Thirdly, they are given chances to sensuously experience all the aesthetic there is in language, math and science, and to discover all the language, math and science that is intrinsic in dance: how dance is done and how it is named.

Fourthly, looking back at the dance activity done, and discussing and naming its features and content, the students' general literacy and language development are enhanced. Again, there is an interplay of disciplines, as the participants handle subject specific terms and notions, in dance and in language, mathematics or science. The different vocabularies are in themselves different ways of looking at the world. At this cognitive and reasoning phase in the method, the students, putting words to the discerned subject matter, help out to formulate definitions and axioms. Sixthly, the content is formalised in sentences, lists of verbs and tenses, equations and formulas, axioms and drawings.

Seventhly and lastly, the aesthetic teaching method caters for additional educative benefits in terms of critical thinking and what is commonly called Bildung or liberal art education: to look at the world in multiple and shifting lights.

While the ambition to invite the student into the world of dance is unabated, in line with the argumentation for aesthetic teaching, the aesthetic experiencing unfolding in these contexts of teaching another subject is less open-ended. The dances 'point' at the embedded subject matter, which renders an intentional aspect to the experience: an aspect of control. In Dewey's (1938/1997) thinking, this is 'organically' connected with the requirement that experiences, in order to be educative, must lead onto an expanded world of subject matter, of facts and ideas. When applied to aesthetic experience, the aesthetic is at risk of being reduced, if not carefully handled and preserved in a deliberate method. Dewey calls for a systematic application of scientific method to ideally benefit from the potentialities inherent in experience in education. Here dance establishes a *space of learning* as coined by professor of education Ference Marton (2004): an interactive space between teacher and pupil, between pupils, and between the pupil and the subject content—another name for the didactic triangle.

In this aesthetic teaching method concept the participants dance and open their senses onto the other field of knowledge, part of the natural and social world and the world of ideas. The kinaesthetic experience precedes cognition, lowering thresholds to take in subject matter which many perceive as difficult. Dance, with its qualities of corporeal, sensuous and dialogic action, shapes educational relationships. Engaging in the dances, the children and students open themselves, as subject-bodies, to each other, to the music, to the props, and to the subject content emerging in the dances.

Dancing and then discussing, they take initiative in action and speech. They scrutinise their actions and ways of reasoning, their previous knowledge and conceptualisations in the subject. Sensuously interacting and playfully exploring, they give way for the yet unknown. They help out to take apart the subject content inherent in the dances. Naming it in appropriate language, they become conscious of the content, explore it deeper, and put it together into a mastered whole.

The result is wonder and enlightenment, characteristic of the consummation of an aesthetic experience. Dance emerges as the archetypical form of teaching, to point: 'Look, there!' Together the young dancers move, sense, discover, consider and figure out—letting dance teach the subject.

Part II
Teaching Dance: A Method

Teaching young people dance is a most rewarding and inspiring task. To experience the power and joy of dancing together, exploring movement, improvising and choreographing, witnessing how much dance a group can charm out of each other, is enlightening as to what being in school can be like.

In order to guide young people through the world of dance, you do not need to be a trained dancer or a schooled dance pedagogue. As Susan Stinson underscores: what you need is an insight in the special nature of dance and its basic elements, and a good methodical foundation to build and further develop your dance teaching on. These insights and proficiencies can only be acquired by dancing—literally step by step. So dance, and let dance show the way.

Chapter 2
The Nature and Elements of Dance

P. Moerman, *A K-12 Dance Method to Teach Creative Dance, English, Math and Science*, https://doi.org/10.1007/978-3-031-93156-7_2

What Dance Is

What is dance? What is dance about? What is the difference between movement and dance?

The short and simple answer to the first question is: dance is movement. It is also the answer to the second question: dance *is* movement; it is not *about* something, other than about *dancing* the movement. This goes for the professional dancer as for anyone working with dance in education. Dance is an art mode and a mode of knowledge, two veins from the same wellspring. That is how dance as art and subject must be treated in all school settings.

All movement can be turned into dance. To walk, run, jump, turn, sneak on the tips of your toes, lift or wave an arm or a leg or a little finger, raise your eyebrows. To heave your chest, pout your lips, pucker your forehead, twist your neck, fly out a knee, fall down, crawl back, roll over.

Movement is also coming to a halt, to stand, stand tall, stand on your toes, go down on your knees, balance on one leg, shiver, shake, skip away, rush headlong faster and faster, spin around, hurl back, collapse, sit down on the floor, on a chair, on a rock by the sea or a boulder in the forest.

Varying the flow of movement in space and time, with different inputs of body and force, adds multiple layers and colours to how a movement can be charged and perceived. By probing and exploring these basics of dance and possibilities of variation, the dancer moves from mere physical gesture to dance. To dance quickly or slowly, softly or shaky, big or little, with one, several or all body parts at the time, on one and the same spot or all over the floor, with full strength or holding back force, in ten brisk repetitions or in an endlessly drawn out gesture.

Kinaesthetic Awareness: Sensing Movement

The purpose of dance education is first and foremost to offer the young dancer a richness of opportunities to become *kinaesthetically aware*: to get a sense of movement, a feeling for movement. Kinaesthetic awareness is what turns movement into dance and is essential in order to perceive dance as an art mode. It may be repeated: *kinesthesia*, from Greek *kinéin* + *aisthesis*, is as much about initiating as sensing movement. It underscores as much the ability to act as the capacity to sensuously perceive movement, one's own or a co-dancer's, from a dancer on stage or in a dance film. It concerns the capacity to be affected, to be *moved* by movement. Kinaesthetic experience evolves by doing and undergoing movement, to go along with a movement to where it takes one, in a sensuous dance dialogue without given outcome.

Think of a young child in a room in preschool or in a playground, wholeheartedly indulged in dance, giving free rein to imagination, taking new initiatives every moment. The child *dances* every movement, is absorbed by movement, as if vanished into an own universe—a world of dance. But that world is for real, it is the real world of space and time. Moreso, with every move the child *enters* the world, finding a place in it and making it their own. What we witness is far more than a physical body in motion. The child does not engage in mere mechanical activity performed by a biological apparatus of movement. Wholly devoted to dance and play, the child moves beyond the predetermined and past all that has been before, breaking into the world anew as a subject-body. The child and their dance are one, by force of imagination, power of initiative and the ability to surrender to movement, a blend of activity and passivity.

In this lies the very nature and meaning of dance. Dance, says Stinson, is about making movement meaningful in itself, by *attending* to it. This goes for any movement, even those we perform perhaps unheedingly every day. We attend to a movement by sensuously experiencing it. In dance, meaning of movement is created in the dancer's capability to kinaesthetically experience it and in the beholder's capability to perceive and experience it in turn. To dance is to discover a magic world of movement. Does dance need to be more?

Consequently, Stinson emphasises that teaching dance first of all must perceive the child or pupil as creator and choreographer rather than as interpreter of demonstrated movements and ideas. Dance instruction must promote play and improvisation rather than imitation. It must aim at the young dancers' discovery and exploration of movement's potential and richness of variation, in joint interplay. Teaching dance must be directed at the young dancers as subjects, providing space to bring their initiatives to the world.

Expression?

Children dancing fully present find a self-evident meaning in their dance. The professional dancer and choreographer has the same driving power to surrender to dance, to try out movement and find meaning in itself while doing it. This is not the same as saying that the child or the dancer is clear about what they want to 'say' or 'express' in their dance. To claim that a dance movement or sequence is 'embodied thought' is anachronic. Kinaesthetic awareness is about sensing movement *before* thought or intention. It makes movement larger, lither and more agile than thought. Dance precedes thought, dance can be *forethought*: choreographer William Forsythe (in Manning & Massumi, 2014) warns his dancers against being too *forewise* in trying out a movement, and so impede its force and potential meaning. Movement gives way to movement and embodies nothing besides itself. As stated before, Manning (2009) views every new movement as body and dance becoming.

Does anything need to be 'expressed' in dance at all? After many years of teaching dance, Stinson (2016) scrutinised her work and came to the conclusion that she had not really explored the wonderful world of dance with the children. She had taught them to be *like* something. Like seeds growing to be flowers, wild horses galloping, ice cream melting. She wondered why children only danced dances that told stories, why they were so *literary*. They *pretended* to be someone or something, they danced emotions. They did doll dances, witch dances, storm dances, battle dances, animal dances, dances about airplanes and dragons, about women cleaning. Why didn't they dance dances that were *dances*? Why did they busy themselves with *ideas* rather than with *movement*'s potential for ideas?

Stinson realised she had circumvented the art of dance. She had taught the stuff dance can relate to but was not dance *in itself*—the eternal issue of the claim on the arts' causal value. Stinson shifted to talk about movement only, without focus on fiction. Together with the children she now went for the experience of flinging out with an arm or boxing with the knees. It caught on among the children and proved to be at least as exciting as wild horses. The difference was that the children now learned about dance and movement and about themselves and their bodies. Now it was for real, not fabricated. Now they could let themselves go and fantasise about Ms. Knee and Mr. Elbow flying up in the air, and the children got to learn more about dance than they ever had through images of flying witches and cleaning moms. Or Alfie Atkins' apron wearing dad.

The primary purpose of dance instruction is to make children just *dance*, to help them become kinaesthetically aware, of shape, force, speed and other aspects of movement, and deepen that sensation. To let them fill themselves with movement, and guide them to experience dance stripped from everything burdening it—not pretending and not refraining from *attending* to movement, that fundamental quality that turns it into dance. As the young dancers come to insight what dance can embrace, acquaint themselves with its basics, having explored and experimented with the craft of dance, they will tie dance to other fields with the help of their imagination. This can be the next step, never the first, Stinson insists.

If fantasy and metaphor lead on to a deeper awareness of movement, movement is the first consideration. Then images can be helpful, such as 'Make your body round like an orange', 'Gallop like a wild horse', 'Cup your hand softly, turn out your wrist in a little circle as if you gently picked a mature orange' or 'Feel the turmoil whirling from within and break out in anger'. In the same manner pictures *arising* from a movement may help, such as 'What is there that moves like that?' Stinson argues it does not help, especially in the beginning of dance instruction, to make children dance oranges, horses or anger. It risks creating trip-wires or locking dance and movement into clichés before the children start getting a grip of the essence of dance and learn how to play with its building blocks.

The point made here is not about banning imageable connections like to galloping horses. It is about not confining the sensation of dancing at a gallop by associating this with the world of animals or physically imitating how horses move—which of course is impossible, unless you are a centaur. It is above all about avoiding reproducing clichés about how an angry or sad person 'is' or 'looks like' or moves. When reduced to a cliché the angry child is objectified; in the worst case the child is made into a caricature of a living person who has all the reason in the world to be angry. How *this* angry child and *this* teenager flying into a tantrum dance their rage is how they are seen as unique dancing individuals and are given a chance to come into being as subjects.

Dance teaching cannot do more than show a pathway into the world of dance and movement. By deepening their kinaesthetic awareness, broadening their repertoire and boosting their imagination of movement, trying out new ways and solutions on the floor, the young dancer will inevitably find ways to charge movement with lived experience, character and temper. But whatever their dance expresses in the end is up to them.

The limits to expression in dance, and for the value and relevance of creativity at large, as Biesta (2021) argues, are revealed as the creative expression meets 'the world': where other wills and desires manifest themselves and are dealt with by those who do not claim to occupy the world's centre. What dance teaching can show is that no one is alone in the child group, in the class and in the world. We can let the world speak to us and be on speaking terms with the world, each one in their own unique way, no one taking up space at the expense of someone else.

Playing and improvising without a given outcome, like dancers do, is a big part of the relevance and indispensability of dance in education. Space is created for all to dance and move ahead through play and risk-taking, without necessarily knowing what comes out of the dance, without even knowing beforehand what one may 'learn' from it all—which also touches the issue of how an 'achievement' is to be assessed and granted a grade. This distinguishes teaching dance, art, music, drama or slojd from other subject teaching.

The Building Blocks of Dance

To every practical or aesthetic subject there is a set of basic elements to get familiar with, explore and master step by step. The advantage of dance is that the very youngest explore and try to master movement as soon as they come to the world. The infant searches with their gaze, splays their fingers, tries little by little to roll over, balance the head, sit up, grip a block, pick up a peg and shake it with genuine pride and joy. The child's movements arise in dialogue with the surrounding world—applauded. The gaze and hands are directed at others' gazes and hands, the whole face and body respond to a smile. Nowhere is the *sensing sensed* body-subject more manifest, taking in the world, giving movement meaning, perceiving, receiving and expressing.

So the child starts moving about in the world, in small leaps between the self and the other—an a*esthetic experiencing* which is described both within developmental psychology and, in this book, in educational, aesthetic and ethical, relational and existential terms. Dancing in the playroom or the playground, the child becomes more and more aware of movement and its explorative force, and continues to probe their reach out to the world. It is for the dance teacher to welcome this flow and receptiveness as a gift and point out particularities of movement for greater kinaesthetic awareness.

Aspects of Movement

When mapping and analysing movement, four parameters compose the starting point: *body*, *space*, *force* and *time*. The four aspects are derived from choreographer and dance theorist Rudolph Laban's basic factors in his model for movement analysis—flow, weight, embodied time and space. They are widely used and adapted, also by Stinson (1988), who adds the aspect *relationships*.

The *body*'s movements and displacements are: to walk, run, hop, leap skip, gallop, creep, sneak, turn, roll and fall. The most elementary, mere motoric movements are to *bend*, *stretch* and *twist*. They can be performed with the whole body or with separate body parts: fingers, wrists, arms, the torso, the back, hips, feet and toes. Even the smallest body parts, the eyes, the eyelids, the lips, the forehead, all parts of the face, even the hair, can engage in dance.

In *space* we discern movements' directions and trajectories: in straight tracks or smooth bends, in a circle or in a square, zigzag, alternating right-left, plane, forward, backwards, sideways or slant forward, in front of or behind each other, criss-cross all over the floor or concentrated on one spot. We also discern how movements are done spatially: up and down, low over the floor, high up towards the ceiling, in a corner, along a wall. The choices of movement *shape* the space, in the same manner as a group of dancers shape space as they gather on the smallest possible area, spread out, dance in rows, in one or multiple loops, and so on.

Force of movement comes out of the work of the muscles, increasing or diminishing between tension and relaxation, soft and hard, light and heavy, caressing and throbbing, flat and fiery.

Finally the length of *time* a movement lasts offers a rich range of possibilities in the shifts between fast and slow, momentary and sustained, repetitive or without a determined rhythm, suddenly halting or slowly ceasing and coming to a standstill.

In combination, these four aspects of movement already offer inexhaustible possibilities of variation to try out differences and extremes as to how much body is activated, how space is used, different time aspects, shifts in dynamics and energy inputs.

Qualities of Movement

Over these four parameters another grid of movement terms can be laid to further describe and analyse movement: static or dynamic *balance*, *coordination* between body parts or with the music, *flexibility* in joints and muscles, muscular *strength*, *speed*, *endurance* and *motoric reactivity*.

The mapping of movement aspects and terms leads to six categories of movement quality: *driven movements* in a continuous flow, using more or less muscle force in various tempi; *stroke movements* like smaller or bigger outbursts with small stops; *vibrations* in some body part resulting from a strong muscle tension; *fall movements* as the body surrenders to gravity, which can be slowed down by an upward force equal to gravitation pulling down; a *swing* like a pendulum movement down and up, going along with or resisting gravitation; and *suspended movements* in moments of stillness, when the motion energy stalls and the movement becomes pending, as in the summits of the swing just before the latent energy is released into a falling movement again.

To the body, space, time and force aspects of movement Stinson adds a *relational* dimension. It concerns how these aspects work alongside, and, in extension, how movement is created and charged as dancers interplay with each other.

Teaching Creative Dance

On a micro level, dance teaching concerns helping the child or pupil discover and experience delight even in the simplest of movements: to slowly move the arm in some direction, to slightly twist the head and glance over the shoulder, fully present in every tiny muscle. The ambition is to give the young dancer the chance to experience the sensation of every movement and thereby find the meaning of movement in itself—find the *magic* in movement, as Stinson's (1988) young dancers used to word it. It is about finding the very meaning in and of dance, the fabric dance is woven of.

Always think of the building blocks of dance throughout a dance session, from assembly and warm-up to reassembly and relaxation. Charge the beginning and the end with a moment of stillness and presence. Use greetings, like: 'So now ... let's dance!' or 'For all that dance today: thank you!'

Be specific in your instructions and use an appropriate, straightforward and varied dance language. Vocabulary is an important teaching tool, it helps the dancer to put words to movement, further enhancing kinaesthetic awareness. Modulate your voice when instructing: rhythmise, sing, whisper, hush, rumble, speak while breathing in and out. Reinforce with handclaps, drums, finger cymbals, castanets and other instruments and helping tools. Leave space between the activities for breathers and open up for comments, questions, suggestions for fill-ins and ideas that can inspire the next step or a deepening exercise. Instruct *movement*: what the young dancer tangibly can sense and experience as cooperating aspects of body, force, space, time and interplay.

Body, Force, Space, Time, Relationships

The dancer handles the elements of dance in every singular movement. An arm movement may be minimalistic while the dancer stands still on one spot, or sweeping in a huge curve through the entire room, the dancer leaning forward as if hurrying along. The arm may move isolated while the rest of the body is fixed, or while the upper body twists in the opposite direction and the feet trip on tiptoe. Arm movements may be fast and repetitive or languidly slow like in a dreamy sequence, staccato as if every cell in the body bustles and quivers, or stock-still as if the body got stuck and the arms move stickily as if working through dough or thick honey. An arm may point at a companion dancer or imitate someone else's movements with new nuances of time and space, force and body.

The instant the dancer is about to initiate a movement, body, space, time and force are in interplay. The dance teacher's task is to direct attention to this interplay in the big and the small, and to encourage the young dancer to try different ways to vary their movements: enlarging one aspect in some direction or toning it down in another; testing extremes between forceful, quick sequences and decelerations into freezing in a pose. Leave space for each dancer's imagination and freedom of choice of movement.

Body and Steps

Name *body parts* and *body movements* in simple, straight instructions, naming the elements of dance.

> Run with silent feet, criss-cross over the floor. Sneak on the tips of your toes. Creep with your whole body. Stand on one leg, keep your balance. Sit down in a circle, sit cross-legged like a tailor.

Widen the spectrum from body parts such as head, shoulders, knee and toe to wrist, chest, elbow, back of the head and small of the back.

> Bob your head. Roll your shoulders. Wave your hands. Clench your fists. Twist your wrists, make small circles outward, inward. Go around two by two, walk arm-in-arm. Sway your hips. Stick your elbows out. Put a hand on your breast, feel the beat of your heart. Breathe in through your nose, pull the air into your lungs, expand your chest. Blow out the air through your mouth in quick puffs, pull in your stomach for each puff. Flex your biceps. Point with the tip of your toe. Flex your foot. Plant the ball of your foot down with a muffled sound. Shake your whole body. Stretch your back. Fall down on your side. Lie still on your side. Flip over to the other side. Roll back. Rest on your back. Flatten down the curve of your back. Lie on your stomach. Worm your way over the floor. Spin around on your bum. Close your eyes. Wrinkle your forehead. Stare straight ahead. Relax your hairline. Keep your eyes closed, relax your temples.

The body's original movements *stretch*, *bend* and *twist*—or *extension*, *flexion* and *rotation*—occur in all the exercises described. They can be accomplished according to what movement the joints allow. Shoulders and hips manage all three, while the elbow and the knee cannot twist. Stretching and bending are each other's opposites, as the joints open and close. Twists and turns are both rotations, the former limited in every body part as no part can twist one whole turn round; only the whole body can do so and then we speak about a *spin* or a *pirouette*. Even *rolling* is rotation, as is *cartwheeling*, *somersaulting* and *flipping*. Stretching and bending evolve also in the opposite pairs *press* and *pull*, as the muscles handle resistance.

As said, the lucky thing about the original movements is that children busy themselves exploring them as soon as they come to the world. Their eagerness to widen their range of movement is a delight to any dance teacher. Help them to enhance their awareness of these elements in simple exercises.

> Sit down in a ring, legs crossed, hands on your knees. Close your eyes. Bow your head slowly, feel the stretch in the back of your neck and the bend in your throat. Raise your head, bend the back of your head backwards, feel the bend in the back of your neck and the stretch in the front. Turn your head to the side, your ear searches your shoulder. Feel the one side of your neck pressing together while the other side stretches out. Go on turning your head down and to the front and on to the other side, all the way back again.

> Straighten up your neck. Open your eyes. Open your eyes wide. Close your eyes again, hard, harder, a little harder still. Open your eyes. Turn your head to one side, to the other side, straight ahead again. Close your eyes. Clench your fists. Unclench, stretch out your fingers. Lift your hands from your knees. Reach out your hands straight ahead in front of your body, the back of your hands to the ceiling. Turn your straight arms and hands with the palm of your hands to the ceiling, feel the twist in your shoulders. Stretch out your arms still longer from the shoulders all the way to the tips of your fingers, feel the energy reach out in the room, straighten your back, feel the force in your body. Cup your hands, pull in your arms, as if you drew in an invisible rope, elbows to the sides of the body. Open your hands, push your arms forward again, as if you pushed away an invisible wall. Shake your fingers, shake your hands, shake your arms, elbows, shoulders. Shake your whole body. Relax all your joints, put your hands back on your knees.

Let them explore further, ask them to dance with their hands, with their toes, in silence or to music. Ask them to dance with different body parts, small and big, sitting, standing or lying down, with small and big movements. Pick up their suggestions, let everybody try out ideas and find new variations.

A number of basic *steps* also occur in the dance exercises further on. All of them are *displacement steps* with the feet, performed in space and time, that is, in directions and tempi. Ask the participants to show which steps there are to get from one place to another, over the floor, up in the air, or striding around on the spot—try them out together. Ask what the steps are called. Summarise the basic steps of dance:

> *Walk*—a weight transfer from one foot over to the other, one foot at the time keeping contact with the floor; at a certain pace.

> *Run*—a weight transfer from one foot over to the other, the body shortly lifting from the floor; at a faster pace.

> *Leap*—an extended run, the body staying suspended in the air for a little while; at an even or uneven pace.

> *Jump*—running or standing, a weight transfer taking off and landing on both feet; at an even or uneven pace.

> *Hop*—a weight transfer taking off and landing on the same foot; at an even or uneven pace.

> *Slide*—a walk at an uneven pace, the front foot sliding over the floor.

> *Gallop*—a combination of a step and a leap, one foot a bit in front of the other, in an uneven 6/8-rhythm, the back foot taking off on the upbeat, the front foot landing on the beat: *gall-op! … gall-op! … gall-op! …*

Skip—a combination of a step and a hop, alternating feet, in an uneven 6/8-rhythm, the same foot taking off on the beat and bouncing on the upbeat to the next beat and foot: *skip! ... a-skip! ...a-skip! ... a-*.

The smart thing about most of these basic steps, as their descriptions take a few words and may appear knotty in text, is that they hardly need to be taught. Many young children gallop and skip, at home, in the playground or in preschool, without anyone having instructed them to 'learn' the steps. The playing child's good temper and joy of imagination make them set off galloping or skipping about. This is just like the infant who fully concentrated tests and tries to master new movements, rolls over or suddenly succeeds in sitting up, to the parents' great joy and pride. As most children are eager to expand their scope of movement, it is a welcome task for the dance educator to work through the skipping or galloping step if coordination seems a bit of a struggle for a child. That is not to say that dance does not need to be taught, or that dance would be an innate or 'natural' skill just like that. So pep up for variation and deepening into all steps:

Gallop forward to the middle of the room and then backwards again. Put your hands on your hips and dance side-galloping all the way to the other side of the room. Gallop lifting your knees high, on the spot, round, round, arms akimbo, up in the air like a V for Victoria. Do four gallop steps with one foot ahead, switch and gallop four times with the other foot ahead. Skip two at a time, one from each side of the room towards each other, get arm-in-arm and skip around two rounds, detach and skip on to another friend, get arm-in-arm, dance two rounds and skip away to the next.

One trick to sort out the *skip* step is to start from gallop and show how skip simply is a variation on gallop. Pardon? Yes:

Gallop four times with one leg ahead. Change leg, without falling out of the rhythm, do four times with the other leg ahead. Once again, four times with one leg, and four times with the other leg ahead.

Do this a couple of times with sustained rhythm.

Continue galloping, now two times with one leg ahead, change, twice with the other leg.

Do it a few times again, so you get the rhythm.

Finally, gallop alternating one time with each leg ahead. Which step do you get now … ? Do you skip …? Yes, now you're skipping!

Sum up, clarifying that a *skip* is a *gallop* with continuous leg shifts.

Go through these basic displacement steps, dancing and talking, and clarify that just like *skip* is a variation on *gallop*, both steps are variations on a *run*. A *run* in turn is a quick variation of a *walk*, as a *leap* and a *glide* are basically *walking* steps.

In summary, the body's displacement steps on the feet can roughly be divided in *walks* and *leaps*, the latter with different variations according to which foot takes off or lands.

Besides displacement steps, by which the body takes in floor and space and which to a degree put demands on balance, there are endless movements that can be done on the spot, such as *shakes* with single body parts or the whole body, rough *jerks*, slack *waggles* or quick *shivers*.

We work with balance as we start to move. Actually already while we stand, or sit. Simple exercises like shifting weight from one leg to the other while standing, or from one buttock to the other while sitting, help young dancers to become aware of this intricate interplay. Someone who is on the *go* or in a *run* and loses balance *stumbles*. It is so much fun then to play around with balance in a dance exercise where you check how much you can stretch and stretch or lean and lean in one direction, give in to gravity and *fall* in different ways—without hurting: on soft body parts, or rolling or going down body part by body part. *Swing* is still another way to play with gravity and let go: to hold on to a state of equilibrium between body and gravitation, to go along with the falling movement again, with the whole body tripping over the floor, or just with an arm, a leg, the torso, the head. A greater control of balance, by holding back in the swings in suspension, gives rise to a *sway*.

Further body-space aspects to explore are *rising* and *lowering* the body, *raising* and *sinking* an arm, an elbow, the chest or a hip, *up* and *down*, often sequentially. Explore shifting between different levels, *low down* over the floor, *high up* over the floor. Similarly investigate shifts between *forward*—'where your nose or your front points'—and *backwards*—'where your back leads'—*slant forward* or *slant backwards, in a zigzag, to the one side, to the other side*. When instructing the youngest children, speaking about *right* and *left* is uncalled for. Introduce the notions in some exercise, mention them during a talk, or wait until a child brings this up.

Explore alternating small and big walking and running steps, leaping, hopping, galloping and skipping, in different directions. Keep in mind that sweeping and energetic movements may be easier for young children. Let them try to vary and lessen the scope of movements, or hold back and then enlarge again. Make use of drawings and pictures. Show a *line* or *pathway* over the floor with *soft bends* or *angular* lines with *right angles*, or a trajectory in *circles*, *straight* stretches or in a *zigzag*. Make flags together, with the movement words *gallop* and *skip* on them:

> Line up in two rows behind each other. The one loop dances through the room in a gallop, the other coil skipping behind each other.

Space and Form

If we think of complete *stillness*, a body takes up as much space in the room as a body in movement. The difference is that movement, however small, always goes in some direction, takes up *space* differently, is performed at different *spots* in the room or along a certain *track* or *pathway*.

In stillness as well as in movement, the body takes and shifts *form*. One simple way to become aware of form is to 'freeze', in a sudden stop or slowing down to a halt. Or to hold a standing, sitting or lying position and imagine someone comes and picks you up from the floor while you keep the form intact. Weight and balance manifest themselves in the muscles' work to maintain form—charged with an amassed, *potential* energy. Think of a cat standing stock still in the middle of a creeping movement, every muscle, every fibre, from the whiskers to the tip of the tail, is ready to initiate movement, a little twist of the neck or a leap on to the table.

Practice holding a form while sitting:

Sit on the floor, feet together, soles of the feet steady on the floor, knees together pulled up to the chest, hold around your knees with your hands, elbows against the body, chin resting on your knees, back bowed. Keep your body together like a little bundle, try to hold that form all the way through the exercise. Now try to lift the soles of your feet from the floor, balance, put your feet down on the floor again. Lift your feet again, swing softly back and forth, put your feet down. Increase the swinging movement, lean your body backwards and forward, balance. Increase the swing a little more, roll over on the small of your back, then back again, back and forth, roll over all the way on your back, your neck touches the floor, roll back.

Sit still again, as in the beginning. Rock from buttock to buttock, side to side, bigger and bigger, tip over to the side, then back again. Keep rolling over to the side, forward and backwards. Try to roll-dance over the floor like a ball of fluff in the wind or a swirl in the water.

Put on the music and ask the children to move in different ways, either from a standing, a sitting or a lying position. When the music stops or on a beat of a drum they freeze their movement and their form. Go around and lift the children gently in your arms one by one, carry them a bit away and put them down in a row, in a circle or a square or on a diagonal, creating a sculpture park. Draw the children's attention to the magic kinaesthetic effect of the collection of child statues. Help them to describe the shapes and put words to aspects of form: *straight* and *bent* or *curled* lines, *blunt* and *sharp* angles, *flat* or *arched* areas, *wide* and *narrow* forms, *compact* shapes or *hollow* when the body or parts of it enclose air space—*holes*. On the drum beat everyone moves from one frozen form to another, alternating with one body part at the time, several parts or the whole body, freezing and moving on each new beat of the drum or a mark in the music.

Other ways to play with form are to let the children try out doing a board or a stick, a bridge or a vault, ball or tube, standing, sitting, lying. Help them relate their *inner sense* of form and the work of their muscles to an *outer feeling* of the room as they take and shift *positions*, *high* or *low*, *against the walls* or *in the middle*. Stillness and movement presuppose each other. Let them try out shifting between stillness and restrained or frenetic movements.

Stand straight, arms down along the body, look straight ahead. Straighten your back, breath in, raise your chest, just a little, lift your chin. Look up, see the ceiling, let your face follow your gaze, lift your hands along your sides, up and past your shoulders, neck, head, your hands aim up towards the ceiling where your eyes are looking. Look down again, curl down, crumple into a little mite, crouch … tumble and fall down.

Time and Rhythm

When we speak about *fast* and *slow*, *rhythm* and *pulse*, we are into *time aspects* of movement and movement sequences. If we go into *accents* in the rhythm, we also take in the *force aspect*, as the intensity of a tone is marked through the force in movement. Here we deal with rhythm aspects such as *accent* or *stress*, *pulse*, *tempo* or *basic rhythm*, *pattern* and *phrasing*, in tune with how the music handles these building blocks of its own—though later on we will see how we can dance *against* the music, contrast it or dance independently from the music, staying with our own pulse or beat.

The most essential of rhythms is the organic: the heartbeat or the breathing rate. Human rhythms are connected to nature's, to the changes of the seasons and shifts in the weather, the eternal pulsations of winds, waves and rapids, the organic movements of animals, birds and insects. Dewey relates all aesthetic experience and all artistic creativity to the rhythms of nature, referring to how humans always have created their rituals interplaying with nature's pulses and the earth's movements.

Besides rhythm, every movement also has *speed* and *duration*. Even *repetition* defines rhythm, as does *deceleration* or *ritardando*, *pauses* and *stops*. Together, *rhythm*, *speed* and *duration* constitute a movement's rhythmical *pattern*.

Vary music and the use of instruments between rhythmical or metrical in different tempi, and in a time of its own with a free flow of movement. Ask the children to move to different paces and genres, ask them how the tunes *sound in their ears* and *feel inside* during a dance. Keep in mind that *fast* to younger children can be a great deal faster than to adults. In the same manner, as shown below in a controlled *use of force*, dancing a whole sequence in *slow motion* can be a special kinaesthetic sensation. Instruct and create rhythmical patterns and phrases by using your voice. Remember the overall purpose is to enhance awareness of all these rhythmic aspects of movement, not to do it 'right' or to try to reproduce. Allow yourself to be surprised by the children's solutions.

> Triple time: And *swing it* and *swing it* and *swing it* ...! Walz: *Step* toe toe, *step* toe toe, *step* toe toe.

Force, Weight and Flow

A movement can be charged with *force*: according to how much the dancer *tenses* or *relaxes* their muscles, *takes off* or *holds back*, making the movement *frenzied* and *sharp* or *soft* and *restrained*; according to how much *strength* the movement is given, *hard* or *loose*; according to how much *weight* the dancer puts in making the movement *heavy* or *light*; or according to how much *flow* is created in the movement, *free*, *bound* or in *balance*. Force and weight constitute the work of *the muscles*, which originates movement and also impacts on *breathing*.

Explore again shifts between *tension* and *relaxation*. In their extremes, both result in the absence of movement. Both are therefore worthwhile examining for a greater awareness of the significance and potential of force for all movement.

> Clench your fists, contract your stomach, press your eyes together, pull your whole body together hard, hold on as long as you can, and feel from within how all that crammed force just wants to burst out in movement.

Releasing tension altogether now and then, especially at the conclusion of the dance session, gives the young dancer opportunities to learn to master their force.

Explore ways of handling weight:

> Walk or trip helter-skelter over the floor, as feather light and soundless as you can, with your whole body's weight in balance, your arms, legs, torso, head, as if your body and the air around you are in balance. Put more weight into your movements, making your way through the air more forcefully, as if the air were thick as honey. Walk as if on moist and dense sand on a beach against the wind. Walk as if in sand so soft and fine it squeaks as your feet sink down in it.

Relationships

The building blocks of dance body, space, time and force, and how they are apportioned, compose all movement. Also within each of these basic elements the various aspects of movement touched upon so far occur and interplay, in countless variations. This way in every move the dancer creates dynamic relations in the material of movement: between body parts, rhythms, the efforts of the muscles, the use of space, and not least, responding to others' dance movements.

In a broader perspective, this is how children engage in discovering their surroundings and have done so all their lives, literally by moving and expanding their room for manoeuvre. Through sensuous, kinaesthetic experiencing, they seek a dialogue with the social and material world.

Teaching dance in preschool or school ultimately aims at deepening, refining and expanding this ongoing movement on the floor, and is a model of what school is about in a broad sense Teaching dance as initiating into the world and art of dance is an invitation to a specific mode of knowledge and proficiencies, and as such a specific way of dialoguing with the world: a way of *moving* through the world and finding one's place there, among the others and what is other, encountering the unfamiliar and the unknown, and encountering oneself in all this. It is an invitation to let dance show the way, to let dance teach us what it is to be in dialogue with the world.

Helping Tools

In many of the exercises and lessons described below, props are used to reinforce movement: flags and pennants, fabrics, tambourines, pictures. The focal point in all activities is the young dancer's sense and imagination of movement, their kinaesthetic awareness and how they can train and expand their dancing skills—with no other input but the young dancers themselves, their individual dance resources and what they can accomplish putting it all together on an empty floor.

Holding on and relating to a prop, like a long slim pennant or a piece of fabric, helps the child to further experience a movement's form and rhythm, extend its spatial flow, or charge the prop itself with movement force. The kinaesthetic experience is rendered an additional visual, auditory or tactile quality. A child can move focus from their body to handling the prop, to the benefit of more security. They handle the pennant, the piece of cloth or the tambourine in a dancing way, letting the prop inspire to find new possibilities of movement.

A Dance Session's Setup

To teach dance is to prop ajar the gate to a world of movement and kinaesthetic experiencing. It is an invitation to a world for the pupils to enter into and come into being by force of their imagination, power of initiative and responsivity to movement. Every dance session is an invitation to yet another tour in the wonderful world of dance. The invitation needs to be articulated and attractive for the child or pupil to be willing to participate and enrich the activities with own ideas. Give the session a clear structure with diverse and balanced content across the dance segments. Think of each session as part of a plan over a longer period. The youngest should be able to keep concentration for some three quarters of an hour, the older children and pupils from one to one and a half hours. Plan and arrange for a series of whole days of dance.

Dancing barefoot is best, not least for young children. It provides the best contact with the floor, from which all movement starts—*the floor is your best friend!* The next best is dancing in soft dance shoes, slippers or slip-on socks with anti-slip soles. Socks and shoes are less appropriate; no one wants to fall on their face or step on someone else's toes. Sneakers are appropriate for *street dance*, *hip-hop*, *breakdance*, *locking*, *popping* or *house*.

Prepare a clear structure and be ready to modify and go off course, following the young dancers' responses and interpretations of the material in relation to your intentions with the exercises. Tune into the level of concentration and involvement in the group and if some part seems easier or harder. Slow down if need be, deepen into some moment or transfer onto a new theme. Share briefly before starting what today's dance session will cover and return on the content at the end during a concluding talk: a so-called *meta-talk* on what you've done together and picked up about dance.

The overall purpose of a dance session is to offer the young dancers new opportunities to explore movement, enhance kinaesthetic awareness, increase proficiencies and create dance. Point out the building blocks of dance and give the keys to improvisation and composition work. The focal point of each session is on the young dancer's kinaesthetic awareness and freedom of play and trial.

This freedom is the freedom to act, ultimately to add something to the world: one's own lived world, the school world, the child group, the classroom. Sojourns in the world of dance are opportunities to indulge oneself in dance, to look at the world through the dancer's gaze, to let dance reveal the world and how it opens for one to step into.

Assembly and Warm-Up

Start by gathering. Some children, younger or older, come in with great zest and gaiety, bubbling with life and eager to get going; some literally fling themselves over the floor like dice; some shadowbox or do breakdance moves. Others seem shy and expectant, refuse flatly to join in or stiffen with fear and press against the wall. Invite them all to join in. Let those who are way too uncomfortable just sit and watch and hop in after a while, or next time. Ask them for a comment now and then. Or for a suggestion for yet another gait or another way to lift an arm.

Some need to warm up, some just need to collect energy. The youngest ones have soft joints, the older ones may need to limber up. Below are a number of dance activities to initiate a session with.

An appropriate gathering formation is to stand or sit in a circle, either in a small intimate ring or as big as the room allows. Some other time, let everybody sit or lie down, spread out on the floor, all still and quiet. To the first notes in the music, a beat on the tambourine or a pling on the triangle the dancers slowly get up, body part by body part. Or let everybody stand around randomly and wake up body and room by stretching arms and legs in all directions. Or let the whole group enter the room dancing in one, two or three loops behind each other.

The beginning of the dance session marks that now we move into the magic world of dance.

Say Hi with a Movement

Make a circle. This is easiest when everybody holds each other's hands until the circle is formed. Take a step back for more elbow room, or forward when the gaps between everyone are too big.

> Stand hands on hips, heels together, feet making a V. Everyone in turn says hi and their name following a ritual: 'Hi, I am Noa!' Everybody answers: 'Hi Noa!' 'Thank you!' says the one who said hi. They turn to the next one in the circle, with their gaze, arms and hands extending and the words: 'Here you are!', upon which the next one says hi and their name, and everybody answers back. Always return to the starting position after each greeting: a clear and simple bodily form reinforcing the shape of the circle and the sense of space in the room.

When everyone has gotten used to the form and the ritual, marking the beginning of today's dance session, everyone eventually adds a movement to the greeting. The one saying hi says their name: 'Hi, my name is Kim!' and makes a movement of their own choice at the same time. If needed, start with an example:

> 'Hi, I am …' One hand goes from the hip in a wide arch out to the side, up in a circle movement above the head and all the way down to the other hip, then sweeping horizontally in an inviting gesture to all in the circle.. Hand and fingers start flapping like a butterfly, frenetically to and fro in the air, until the other hand grabs it by the wrist and brings it back to the hip '… Paul!'

All respond saying hi and your name while doing the same movement. Then you say 'Thank you!' and 'Here you are!' to the next one in the circle. Encourage everyone to pick the first idea that comes up, it can be the simplest of movements: a wave with the hand, a stamp with the foot, a turn with the body or two. Lots of inventive suggestions come up:

A curtsy. One or several handclaps. A stamp with each foot. A tap on the knee. A turn on the spot, the hands on the hips, on the shoulders or the arms stretching out. Rising on the toes. Going down on the knees and up again. One hop forward and one backwards. A turn with the wheelchair. A wild run all around or in and out of the ring and back to one's place. Someone takes their neighbour's hands or grabs arm-in-arm and spins around, and so forth.

Remind the group about the form for the greeting ritual and the starting position: hands back on the hips and feet in a V after each greeting. The ritual is the dance form and the framework within which everyone is invited to fill in with their own idea. As the personal movement is tried out and reaffirmed by everyone, the young dancer obtains their own place in the community of dancers.

Dancing into the Room in One or Several Loops

Music: 4-beat, driving, e.g. Juan Peña Lebrijano & Orquesta Andalusí de Tanger: El Anillo.

Everybody lines up in one, two or three files, depending on the size of the group and the room. If there are two or more entrances to the room, the lines can be formed outside and come in dancing when the music starts.

The first one in line holds a pennant or a flag with 'Gallop', 'Skip' or 'Slide' on them, dancing through the room in twists and turns as they wish. Those behind each flag holder have their hands on their backs and gallop, skip or slide along. Emphasise the rhythm with a drum or a tambourine.

Develop the dance with ideas of your own or from a child or pupil:

Stop the rhythm by a beat on the drum every now and then to interrupt the flow for a while and insert a new movement. It may be soft and contrasting on the spot; for example, put your hands together high over your head and bring your hands and arms down in a slithering, snake-like movement engaging the whole body, all the way squatting down, then up again like a coiling snake, down and up again a couple of times. Beat the drum, and the flag holders and their loops resume their tracks dancing over the floor.

Beat the drum close to a dancer last in line, who halts, spins around arms extended and sits down. Beat the drum again next to one galloping, skipping or sliding last in line, who spins and sits down. The others continue dancing through the room. Those sitting on the floor clap the rhythm. As more and more sit spread over the floor, the flag holders and their shortening files dance their way between them. Finally the leaders gallop, skip or slide between those sitting, until they halt on a drum beat, spin around flag high and sit down. Everybody claps the rhythm to the end of the music.

In one variation with two loops, the dance finishes side-galloping in two circles. The one leader and their loop dance in an inner ring in one direction, the others in an outer ring in the other direction. Everyone turns their face to the middle, hands to the sides, and the leaders hold their flags. Beat the rhythm on the drum, dance along in the outer ring. Count eight beats, call out 'Turn outside!' Side-galloping, everyone turns their back to the middle, hands up in the air. Count eight beats again, call out 'Turn inside!', everyone turns to the centre again. Count four beats, 'Turn outside!' Four beats, 'Turn inside!' Try two beats out and in again, then one beat at the time: 'Turn out, turn in, turn out, turn in'. Stop the motion. Clap the rhythm all together, then clap and stamp. Clapping and stamping, everybody moves into the middle, closer and closer to each other, then backs out again. Let go the rhythm, everyone spreads out, spins around, arms stretching out and slowly going down through the knees, sitting down legs crossed. Arms do waving, billowing movements, airy and loose, smaller and smaller, in silence after the music has tuned out.

The Whole Body Walking

Music: rhythmic, exciting, powerful, e.g. Bizet, Carmen: overture.

All movement is conditioned by gravity drawing the body to the floor, which puts demands on balance and coordination. *The floor is the dancer's best friend*, so goes the saying in dance pedagogy. A good motoric feeling for the feet's contact with the floor is essential to understanding movement.

We can all walk, we all came to school today, there's nothing to it. Now, we'll walk with the whole body.

Put on the music. Walk, moving your whole body. Arms, elbows, shoulders, head, hair, stomach, hips, legs, knees and toes—make your whole body walk! Go around helter-skelter, change directions, in awkward angles.

At the change of the melody in the music: Walk on your toes, tip, toe, on your toes, tip, toe. Walk on your heels, heels, heels, on your heels, heels, heels. On your toes, on your heels, your toes, your heels. Toes, heels, toes, heels, toes, heels.

At the change of melody: March like a soldier, march, march, march ... Swing those arms, lift those knees. *New change of melody*: Walk on the inside of your feet, on the inside, do you find the inside? Walk on the outside of your feet, outside, outside. On the inside, outside, inside, outside ...

New change of melody: Walk and move like spaghetti, fling about your arms and legs, feet and knees. All your joints are doughy, knees give way, your head flings hither and thither, your hair with it, hips sway all over the place. Move like boiling spaghetti. Walk around with your whole body.

When the music stops: Freeze!

Change of melody: Steal on the tips of your toes ...! *On the timpani beats*: Stretch your arms up! Down! One arm out to the side, the other arm to the side! Quiet feet, fast feet, tripping on tiptoe, trip to the walls. Backs turned to the walls, arms stretched to the sides, keep tripping on tiptoe sideways along the walls, all in the same direction.

Hold on! Steal forward on the tips of your toes, towards the middle of the room. *On the timpani beats in the music*: Arms up! Down! One arm to the side, other arm to the other side! Quiet feet, fast feet, tripping on tiptoe, trip back to the walls. Turn your noses to the walls, arms out to the sides, keep tripping on tiptoes sideways along the walls.

Stop! Steal backwards on the tips of your toes, fill the floor again. Stretch arms up, down, to the sides. Keep your arms stretched out to the sides. *The music rises*: Spin around. Faster, faster, still faster ... *The music stops at once*: Sit down.

Talk about the dance. About walking, casually, with the whole body. About getting every part of the body going. How it feels to walk with the whole sole of the foot touching the floor, or only toes, or heels, on the outside or inside of your feet. How it feels in the rest of the body, steady and heavy or light and balancing. About the feeling when we moved through the room crisscrossing in angular lines, compared to when we tripped along the walls, how big, empty and inviting the floor felt before we stole forward on tiptoe and took in the floor again. Talk about force, the difference between eager, tripping movements and sneaking. About time, how the spins at the end went faster and faster and faster. Point out more space, force, time and bodily aspects, help the dancers put words to everything. Relate to the driving rhythm in the music. Some other time we will check how it feels to dance slowly to fast music and the other way around.

Share your intentions with these dance plays. To find out different ways to touch the floor with the feet. To train gross and fine motor skills, basic coordination, a sense of space and rhythm, and to feel the joyous experience of dancing all together. Ask for more ideas to dance on the *walk* theme. What gaits are there—manners of walking? Ask them to look at how people walk. Ask them to demonstrate or come up with unusual gaits. Help them to put this together in a wacky walking dance.

Warming Up Body Part by Body Part

Everybody stands in a circle, arms hanging along the sides. Warm up all body parts, one by one, starting with the fingers, then hands, wrists and on through the whole body. Explore the qualities of movement for each body part by isolating and moving it individually while the rest of the body is kept as still and fixed as possible. This exercise in bodily awareness starts at the microlevel, training fine motor skills and evolving towards more and more gross motor awareness.

Your fingers start to dance, arms hanging still. Thumbs start, bending and stretching, making small rings. The forefingers hang on—how can your index fingers dance? Indicating, pointing hither and thither? How does a little finger dance? Pinkie blinking and winking: 'Hey, I want to be in this dance too'. Move your fingers faster, flapping, floundering, rubbing each other. Let them climb up in the air, writing squiggles in the air.

Make use of how the children carry out the movements, imitate their kneading, scratching and itching movements. Ask for yet other ways to dance with the fingers.

Shake your fingers, gently. A little faster, and faster ... hold still. Shake your hands, as to shake off water drops from wet fingers. Stretch one arm up, bring it slowly down while shaking hand and fingers, freeze at hip height, the fingers spread out. Same with the other hand. Again, a bit faster, one hand, the other hand, both. Shake! Freeze! Spread out your fingers, as much as you can, feel the force and tension. Shake. Splay, hold. Clench your fists, hard, harder. Shake your fingers again ... clench your fists, hard, knuckles turning white. Loosen up and do soooft, waving, billowing movements.

Continue and explore wrist movements.

> Do a painting movement with one hand, like a brush. Hand and fingers relaxed, the wrist leading. Up and down, up and down, from side to side, diagonally in the air. Hold your hand at chest height and concentrate on only moving your wrist while your forearm, elbow and upper arm hang on and your fingers are relaxed but fixed still in the air.
>
> Hold one arm still in the air in front of you. Pat lightly on the wrist with the other wrist. A few taps on your wrist, on your forearm, on your elbow, your shoulder, your breast, go on tapping in different spots in the air. Play the drums in the air with both wrists. Let your fingers, hands and wrists dance together.

Add elbows, shoulders, chin and neck.

> Can elbows dance? Here come the kids from Elbow Lake, dancing a tactful, delicate dance on the spot, look how sweet they are. See how they get puffed up, elbowing their way ahead. They take over the place, poor shoulders trying hard to hang in and forearms, upper arms, chest, all moving along in the elbowroom.
>
> Stick your chin out: 'Beg your pardon?', pull in your chin in: 'Nooo!', out again: 'What?', in again: 'Nooo!' Out, in, out, in. Same thing: stick your neck out: 'Excuse me?' Pull back: 'Ooops'.

Maybe someone wants to show how to move their head horizontally sideways like in oriental, Thai or Indian dance. A trick that fits into the style:

> Put the palms of your hands together above your head, elbows to the side. Keep your hands, arms and shoulders still, making it easier to isolate your neck and move it from side to side, forward and back. Try just a few millimetres at the time to discover those small neck muscles. End up moving your head a whole circle around.

Go on and add body part to body part moving them in isolation. Dance with your head, hair, different parts of the face, make grimaces, a wry face. Dance with your torso, stomach, hips, behind, thighs, knees, ankles, toes. Isolate each body part and increase its range of movements.

Wrap it up by starting from the beginning. First just the fingers dance, then fingers and wrists, then fingers, wrists and elbows, and so on, until the whole body is dancing.

Pick up on the intensity of movement in the group. Ask everyone to dance like one of them, then like another, one after the other going through the whole group. The last one leading the dance makes their moves smaller and smaller, little by little, until only the fingers or another body part is dancing, a bit lesser again, slower and slower, until all body parts and the whole body become still.

Progressive Dance Exercises

A dance warm-up is as much an interhuman as a sensuous matter. Together the group of young dancers awaken their kinaesthetic senses, warm up their bodies and come into presence in time and space, stepping into the world of dance, part of the social and cultural world. Wide awake in all meanings, a whole world of wonder lies open for exploration and experience.

Dewey views every experience as interconnected to earlier and coming experiences. In aesthetic experiencing and in teaching dance and the arts there is reason to be cautious about thinking in linear terms of fixed goals and stages. The 'body' of young dancers is a living group of receptive and agentive subject-bodies. Let their response to the exercises in turn guide you in your teacherly mission to offer new opportunities to deepen kinaesthetic awareness. Go back and forth between revisiting the building blocks of dance and progressive dance exercises that appear appropriate in the moment.

Arms like Wings to Fly with

Music: free flow, with rhythmical parts, e.g. Radio Tarifa: Canción Sefardí.

Everybody squats down, arms and hands on their backs.

Let's dance and get a feeling of flying. Let's charge our bodies, our legs and arms with force. We start sitting on our heels, arms behind our backs, hands resting on the small of our backs. Gently roll your shoulders, keep your hands and fingers still, straight yet relaxed. Let's try to rise a bit, straighten your knees just a little. Squat down again, let's wait a second, gather force. Rise again, press yourselves up from your feet, through your lower legs, knees, thighs work hard. Stretch your legs, straighten your backs and spread out your arms, lift them all the way over your heads. Rise on the tips of your toes, stretch your bodies all the way up, how tall can you get?

Now make your body as light as a feather, trip along on tiptoe, soundless, on silent feet. Bring your arms down, slightly bent, strong arms. Up again, and down again, like wing strokes. Up, down, slowly, up and down. Arms are strong, the body light, can we manage to fly? Let's try. Strong bodies, smooth and lithe, pliant and supple. Strong arms, like wings. Up and down. Put on some speed, move through the air, in all directions. Hold your arms stretched out, sail on wings. Fly, to and fro, arms up, down.

Stop, spin around, arms reaching out to the sides, turn around to the one side, to the other, feel the air fan under your arms and hands. Stay on the spot. Do big wing beats: arms up, high above your head, stretch your body, rise on the tips of your toes, straight legs. Lower your arms, squat down, stretch up again, and down, up on your toes. Arms like big, strong wings.

Fly away again, hither and thither, in all directions. Fly and flap, forward, backwards, sideways, crisscross between all these flying friends. Sail high up to the sky.

Break off your flight, point one arm down, the other up. Spin round, look down at the earth, the roofs on the houses, the people down there, children on bikes. Change, point the other arm to the sky, look up to the sky, spin, turn … easy to get dizzy …

Big strong arm movements again, up and down, up and down, stretching your bodies and squatting down, stretching and bending your knees. Light bodies, strong arms, like wings.

Let's fly home now. Just a few more rounds, fast, crisscross, swish, swish, like aerialists, forward, backwards, low down over the ground, high up in the clouds, all over lands and sky. Let's land now, squat down. Fold in your arms. Hop up again, a little bound, arms up.

> Squat down, hop up again, down again on your heels. Once more, and again, and one more time. Land, sit on your heels. Fold in your arms, hands on the small of your back again, roll your shoulders, nice and sweet, put your elbows out, let your arms rest on your backs, strong arms, like wings we could fly with.

Talk about how arms and shoulders feel, and thighs. Those muscles did quite some work. Talk about the feet's silent tripping, how light we can make our bodies thanks to the work of the feet, the legs and the thighs. Yet we can also do those huge strong arm movements with all the force from the muscles. Imagine we were weightless and could fly and whiz through the sky. That's what we did! When we dance we can do anything. The flying theme, the dream of being able to fly catches the children's fancy. The dance reoccurs on the 'Air' theme in Chap. 6 on dancing natural science as the birds' dance, along with a dance with swallows sailing under the clouds. For a language teaching variation of the birds' dance see Chap. 4.

Galloping or Skipping Two by Two

Music: rhythmic, 2 or 4-beat, e.g. Radio Tarifa: Rumba Argelina.

Half the group are standing in a row side by side at the one end of the room, the other half on the other side. Everyone faces a partner opposite across the room. Show the dance steps with one of the dancers, or let two of them show them following the instructions:

> Gallop/skip ahead towards each other, hands on the backs, meet in the middle of the room, gallop/skip a couple of turns around each other, continue galloping/skipping over to the other side.

Show it once more, for clarity and so everyone gets back to where they started. Everybody dances two by two at a time, starting at one end of each row. If the group is big, two pairs dance at a time, starting from both ends of the rows. All stand clapping the rhythm.

> When everyone has done the dance two by two, hop into one of the rows. All dance forward at the same time, meet in the middle, wave their hands, shouting 'Bye, bye!' and gallop/skip backwards, hands on backs again. Repeat a few times.

Progressions:

> The pair meeting in the middle grab each other by the arm and gallop/skip arm-in-arm round a couple of times, holding the other arm up at a right angle. Next time they repeat the turns arm-in-arm, detach, spin round, grab each other by the other arm and do a few turns arm-in-arm in the other direction, the outer arm up in the air.

Talk about the feeling when holding the arm up in the air. How it adds more verve, balance and control to the turning movement. Talk about the strong feeling of force in the spins, as if a force tries to drag you away from each other and fling you out from the centre. It is called a centrifugal force, it tries to make the body fly out from the centre. Like the drying drum in the washing machine, spinning and spinning, faster and faster, hurling the water left in the clothes out through the holes of the drum.

Side-gallop. Everybody is standing in their two rows, now faces turned to one long wall. Hands on hips, one shoulder turned to the opposite row. Everybody side-gallops all the way to the other side, the two lines crossing in the middle. Back and forth, a couple of times.

Talk about the last variation, side-galloping. Even though there are so many people in the room and it gets really cramped in the middle, everyone dancing and crossing each other tightly, I've never ever seen any clashes in any dance session through all the years … How come? How did we do? When we dance we simply sense with our own bodies where the others are, don't we? We get a feeling for the space we share. In the middle of all the galloping going on, everyone seems to find a little gap to pass through. All that speed and power coming out of our movements makes us all the more sensitive to our own bodies and our fellow dancers. When we dance we can do all sorts of things together and come very close to each other in many ways.

Galloping and Jumping over the River

Music: 2 or 4-beat, paso doble, circus music; helping tools: long fabric, hand drum, castanets.

Having galloped many times in different ways, we can of course think of galloping horses. To imagine how horses run and jump adds another flavour to movements and steps like galloping and leaping.

Everybody lines up behind each other at one side of the room, hands on the hips. In the middle of the room across the floor lies a long piece of fabric. We imagine it's a river and we are young horses eager to show off how high we can jump over it. Put on the music, reinforce with castanets or a drum.

The first one in the line gallops towards the cloth, jumps over it, keeps on galloping, stops a bit from the opposite wall and turns around, facing the others waiting in line. The next one has taken off galloping just when the first one jumped over the cloth, jumps in turn, gallops further and lines up behind the first. Everyone so fills up and forms a new line behind each other until everyone is standing on the other side of the room.

Gallop and jump back now, one after the other. We can jump a little higher, can't we? Spread out your arms in the air when jumping, like wings, as if you floated in the air a while before you land on the floor. As if you flew over the river, like Pegasus.

Lie down stretched out full length over the fabric so they jump over you, or put some cushions or a mat in the middle.

Progression: gallop one by one, two by two, four by four, eight by eight … everyone:

One by one.

The first dancer in line takes off galloping, jumps over the fabric in the middle, gallops on to the other side of the room, then turns to the left and gallops back, jumps over the fabric again, keeps galloping and halts by the waiting line. The second dancer has taken off galloping as the first one jumped over the fabric, jumps over and gallops on, then turns to the right and gallops back, jumps over the fabric and comes to a halt to the right of the waiting line, facing the first dancer standing on the other side. The third one dances like the first one, the fourth like the second, and so on. As the waiting line of dancers shortens as one after the other gallops away, jumps and gallops back, the two new lines opposite each other lengthen.

Two by two.

When everyone has galloped, jumped and turned back one by one, the two dancers each standing first in the new lines take each other's hand, put the other hand on their hips and take off galloping together, jump over the fabric and gallop on to the other side of the room, split up and gallop back, one to the left and one to the right, jump over the fabric again, gallop and go stand back in their lines. The second pair starts galloping as the first pair jumps over the fabric, then the next pair sets off, and the next, until everyone has galloped and jumped two by two and the two opposite lines are restored.

Four by four, eight by eight, and so forth.

So it goes on, two from each line take each other by the hand, the outer ones put their free hand on their hips. The four take off galloping, jump, gallop on, split up two by two, turn back in the same direction as before and go stand back in the lines. As the four previous jump, the next quartet takes off.

When everyone has danced four by four, four from each side take each other's hand, the outer ones put their free hands on their hips. Now they gallop eight by eight. Then sixteen by sixteen. A couple of times I've had well over thirty-two young dancers and once more than sixty-four!

Grand finale. All at the same time. They stand close side by side in a row across one side of the room, holding hands, those in the ends keep their free hand on their hip. Everybody ready steady, put one leg in front. On your signal they all gallop and jump over the fabric at the same time. Stand by the fabric and check that they all gallop and jump like one single … horse!

Imitate or Invent: Take After or Do your Own Thing

Most of the exercises described so far imply a degree of instruction or *direct mediation* of movement for the young dancer to try out, leaving space to add a personal touch to perform, interpret and vary. They *take after* without doing the *same*, which would be impossible unless everyone were identical in body and subjectivity. Taking after is therefore not desirable in itself. It may be valuable in contexts where the idea is to achieve a compositional effect drawing on the largest possible likeness and unison movements in the dancers, as in synchronised swimming or mass choreography film scenes.

Dance teaching from the first steps aims at offering opportunities for the young dancer to deepen their kinaesthetic awareness and their capability to create in dance—to make choreographies. Deepening insights and skills in the vocabulary of movement, movement quality and movement variation, they become young choreographers.

The demonstrated movements and instructed dance plays are designed as elementary exercises of steps, coordination, balance, body control and muscle training. They may also serve as an introduction to a specific genre or dance style, such as Spanish dance below, or oriental, folk dance and so on. The dance teacher's or participants' previous knowledge and familiarity with a style or tradition are of great value to share.

The next chapter *Dancing with each Other* is about dance teaching as *indirect mediation*, where the teacher shows no steps. The model builds on improvisation and composition work starting from the participants' joint and shared movement material. In the exercises below the young dancers are encouraged to take more and more control of their movements and to master the language to name them.

Come over Dancing! … How?

Half the group ask the other half to come over dancing in a certain way, paying attention to language.

Half the group is standing side by side at one end of the room, the other half is opposite, facing each other. Ask for the names of their group. *The Squirrels! The Horses!*

> The first group call: 'Hi there Squirrels!' The Squirrels answer: 'Hi!' The Horses go: 'Come over here dancing!' The Squirrels respond: 'How?'

'The Horses' put their heads together. Remind them of the dance words we use to describe different steps, and how we name body parts, speed, direction and energy.

> Gallop backwards! Sneak on tiptoe behind each other! On all fours! Walk on your heels in a zigzag! Squat down and walk over! Hop on both legs back and forth! Spin around with your hands on your heads! Hop from one foot to the other without losing balance! Roll like a stick! Run and tumble all the time! Walk on a tightrope! Crawl on your stomachs like through a tunnel! Skip two by two arm-in-arm! Come over tripping three by three back to back! Keep two feet and one hand on the floor! Stiff and jerky like a robot! Slack like spaghetti!

The Squirrels dance over to the Horses. When they get there the Horses wave 'Bye, bye!' and the Squirrels dance home backwards. Then the Squirrels in turn ask the Horses to come over dancing in a certain fashion. They take turns multiple times.

Talk about the dance in dance language. Go through all the variations and details displayed in the dance play. Point out the richness and power of dance language to name them. Discuss the sensation to make others move in a certain manner just by saying the appropriate words.

Spanish-Inspired Dance: Picking Fruit under the Orange Tree

Music: calm melodic, e.g. Gipsy Kings: Passion, or Ravel: Boléro.

Imagine we are sitting against a tree in a warm country. We will pick fruit and put the fruit in a basket. The movements are Spanish school and flamenco inspired.

Everybody is sitting in a big circle, cross-legged, straight backs, hands behinds the backs.

We do an uninterrupted series of gently curved movements, smooth, graceful, symmetric, in a *bound flow*: with rounded and stretched arms and soft-yet- strong, alert hands, in a Spanish or oriental style, opening and closing. Wrists and elbows initiate and lead the arm movements, carefully with a bound force, as if the joints were moving through smooth honey. The upper part of the body follows along, slightly bending forward, stretching up, or gently to the sides, always returning to a straight back.

We do the same sequence four times, with the same *ongoing*, *bound* quality of movement but shifting *muscle force* and *accents*. Each sequence comprises arm movements *upward*, *forward* and *sideways*. First we do the movements softly and gently, then a bit harder and bolder, then tiny little and finally huge and heavy. Modify your voice accordingly, uttering 'like an orange', 'a banana', 'small, small berries' and 'a watermelon' as to echo the sensation in the movements and the body's work. Not in a documentary or illustrative way, but in order to feel the breeze of a dance world where people move just like this.

Sit with straight backs, legs crossed, hands behind, elbows out to the sides. Arms and hands fold out to the sides and in a wide circle over the floor in front of the body, like a basket, the wrists lead the movement. The upper body joins in, slightly bending forward. Hands meet out front, hands and arms cross each other and continue the circle over the floor. Hands cross each other again in front of the stomach, go out to the sides and make a wide arch back and land on the small of the back again, the upper body straightening up again.

Once again, make a big circle over the floor and a big arch back behind your back. The whole sweeping movement sequence does a big figure of eight, a loop over the floor, a loop wide around the hips. The right and left hand, elbow and arm do the same symmetrical movements, outward to the front, inward and back behind. When we've done the eight the second time, we move our hands back to the small of the back, elbows slightly out.

Now we pick fruit and put the fruit into the basket. One hand stays on the back. Lift the other hand straight upward in front of your stomach and face, the elbow and wrist lifting first, arm stretching right up high over your head. Stretch the upper body. Do an elegant twist with your wrist in a Spanish oriental manner and cup your hand as if you gently pick a sun-ripe orange. Lower your hand again, down in front of your face, your stomach, drop the orange into the basket. Put your hand back behind your back while you start moving the other hand and arm, wrist and elbow first, the same movement upward, pick an orange, take down, drop into the basket, your hand back behind your back.

Now the first hand, wrist, elbow and arm do the same movement, this time straight forward from the shoulder, all the way to a stretched arm, the upper body straight, the chest slightly pressing forward. The wrist does a smooth twist outward, the hand cups gently in an outward picking movement, comes back towards the body, drops into the basket, goes back behind the back, as the other hand and arm start moving the same way up and down and back again.

The same again, this time to the sides. The wrist and elbow keep initiating the movement, the arm stretches gently to the side, the upper body slightly leaning to the side. The wrist does an outward twist, the hand cups and picks, comes back to the body, drops the fruit into the basket and moves back behind the back, the other hand and arm do the same to the other side.

Now we repeat everything at double tempo: up, up, front, front, side, side.

Each sequence, first with the one hand and then with the other arm up, forward and to the side, finishes like it started, with a symmetrical 'big eight' movement, two times:

Do the basket, feel behind your back. Again, do the basket, feel behind your back.

Now do the whole sequence with both arms in three more variations. Shift your tone of voice already in 'Do a basket, feel behind your back' as to call forth a new quality of movement. Hand and wrist always lead, moving smoothly and gracefully in the first, upward variation, then you firmly clench your fist in the picking movement—*'pick … a banana!'*—and bring it easy and controlled down to the basket. Charge your voice with a peevish, angry tone culminating in a snatch, an *accent* flavouring the whole movement, from the fist through the wrist, the elbow and the arm all the way to the shoulder:

Do the basket, feel behind your back. Pick a … banana!

Do the harsh variation upward, forward and to the sides, and then at double tempo. In the next variation the movements are tiny little, eager but delicate. Keep elbows and wrists tight into your body, pull in your shoulders, huddle up, do a miniature basket indicating the 'eight' for the basket and the arch around the hips with your index fingers only. Make your voice little like a baby's:

Do a liiittle basket, feeeel behind your back. Pick … small berries, tiny, tiny small berries …

Do the picking movements with both hands, fingers picking fast, freely to the music, some up, some half way up, to the front, to the side, behind, in all directions, eager, close to the body, a little longer here, a little longer there, put each 'berry' quickly in the basket and pick one more and one more. Accelerate, the movements getting fiery and chaotic.

Change style and tone for the last variation. Slow down and lower your voice like a giant's, a dark, drawling bass:

Do a biiig basket, feeel behind your baaacks!

Maximise the movements, symmetrical again and now simultaneously with both arms and hands, this time only upward.

Do a biiig basket, feeel behind your back. Keep your hands straight and tight, fingers together. Elbows, wrists and arms grow big like a giant's, folding out to the sides, upward in a big arch, the upper body stretching up, chest rising, shoulders heaving, hands shoulder-wide apart. Open your hands, fingers spreading and gripping high above your head, a …

> waaatermelon! Take down the waaatermelon, it's heavyyy, bring it down slowly, it passes in front of your face, the melon crashes straight into the basket, necks bowing, heads almost following after down into the basket. Elbows, arms and hands up again, pick another waaatermelon. And another one, this one is so biiig and heavyyy, no way to lift it down in a controlled way, hands and fingers holding hard around the melon as it crashes down, bangs on your head—'Ouch!'—and bounces down into the basket. One last one, arms up, another waaatermelon bangs on your head—'Ouch!'—and down into the basket.

Now the basket is full. Do the basic movements again.

> Do the basket, hands start from behind your back. Do the basket, arms out to the sides and over the floor in front of you. Hands cross each other, continue the movement in front of your stomach, back around your hips like a big figure of eight. One last time, do the basket and now do a big embracing movement around it, take a hold of it, lift it up carefully, oh so heavy it is, put it on your head. Take away your hands, balance the basket, keep up your gaze, hold the palms of your hands together in front of your chest, straight back.

Stamp the Beat with your Feet: Make Jam

Music: rhythmic, upbeat, e.g. Radio Tarifa: Lama bada.

The orange and fruit theme is helpful for a further introduction to rhythms, Spanish, flamenco or just 4-beat, eighth notes, triplets and so on.

We are still tailor sitting in our ring, all still with straight backs, palms together in front of our chests, elbows a bit lifted, eyes glancing up to our basket full of fruit weighing heavy on our heads.

> Keep your hands together, upper bodies still, backs straight, balance your basket. Get up on your feet, work from your waists, keep your balance. Spin around slowly, one turn, small steps. Can your snap your fingers?

Everybody is standing now, facing the inside of the ring.

> Reach out with your arms, lift the basket from your heads, slowly, it's very heavy, lift it down in front of your bellies, bend your backs a little, get ready, raise and throw the basket with all the oranges and bananas and berries and watermelons up in the air. Now all the fruit lies in front of us on the floor.

Let's make jam.

> Hands on hips, straight backs. Stamp your feet. One foot stamps, the other foot stamps. Follow the 4-beat in the music, stress the one: *pam* two three four, *pam* two three four, *pam* two three four, *pam* two three four.
>
> Half the time, double the stamps, still stressing the one: *pam* and pam and *pam* and pam and *pam* and pam and *pam* and pam.
>
> Half the time and double the stamps once again, into eighth notes, keep stressing the one:
>
> *pam*-pam-pam-pam, *pam*-pam-pam-pam, *pam*-pam-pam-pam, *pam*-pam-pam-pam, *pam*-pam-pam-pam.
>
> Try triplets, *pam*-pam-pam, *pam*-pam-pam, *pam*-pam-pam, *pam*-pam-pam.

Want to half again from eighth notes? Stamp twice as fast again?

Let's try! Try stamping sixteenth notes …

Escalate and finish with soaring, rattling stamps, mashing the fruit to jam.

Stamp and move into the ring, lift your hands in an arch from your hips, to the sides and all the way up over your heads: mash, mash, mash, mash. Keep stamping and back out again, take your arms down, hands back on your hips. Repeat a couple of times, in and out, a billowing movement with your arms and upper body while you keep stamping and moving into the ring mashing and then backing out. Stay back in the ring, hands on hips, feet stamping, lower the tempo, quiet down, stiller, almost unheard fast stamps … until all silent.

Picking Oranges, Bananas, Small Berries and Big Watermelons

Music: rhythmic 4-beat, melodical, e.g. Radio Tarifa: Oye China.

The next dance is a progression, a 'standing' variation on the fruit basket dance. It involves several body parts and rhythms, stretching and bending the legs and feet, more muscle work, more coordination and variation of movement, playful imaginary small throws at each other, a general 'taking after' dance. It contains style elements from a Spanish arm school.

Stand in a ring, hands on hips, heels together, feet in a V. Start the music. Stress the beat on the one.

Close your feet. Open. Close. Half the time: open, close, open, close. Add hands and arms, do a horizontal figure of eight at waist height, like an orchestra conductor: open, close, open, close. Speed up. Open, close, open, close, open and close.

Feet together, parallel. Hands on hips. Stretch a leg forward, point your toes on the floor, bring back. Other leg forward, back. Rhythmise: *toe* two three four, *toe* two three four … Double: *toe* and toe and *toe* and toe ... Add an arm, same as the leg stretching forward, stretch your arm forward in a presenting gesture slant forward, palm of the hand upward: a *toe*, a *toe*, a *toe*, a *toe* ... Feet back together, hands on hips.

Put a heel on the floor, leg stretched forward, back again. Other heel on the floor, back. Rhythmise: *heel* two three four, *heel* two three four ... Double: *heel* and heel and *heel* and heel ... Add arms: stick out parallel arms, clenched fists, shoulder-high. Do a choppy hushing *sh, sh, sh, sh*. Feet back together, hands on hips.

Hop forward on one foot on the *one* … take three steps back, the back foot first … two three four. Hop forward on the other foot, *one* … step back … two three four. Rhythmise: *hop*, go, go, go, *hop*, go, go, go, *hop*, go, go, go, *hop*, go, go, go. Add arms to the hop, flinging them out slant forward upward like a victory gesture. Hands back on the hips on the steps back. Feet back together, hands on hips.

Picking oranges gently; up, front, to the sides.

Parallel feet. Do a basket in front of your belly, each arm making a half circle, hands cross each other up front, continue the curve, hands cross again in front of your belly, continue out to the sides, a wide circle around your hips, behind your back. Symmetrical arm movements describe a horizontal figure of eight.

Do the basket again, hands cross each other and land back on the hips, a horizontal eight.

Lift one arm, elbow first, then the wrist, up in front of your face, hands and fingers hanging relaxed, unfold your arm until stretched. Lift your eyes. High above your head your wrist and hand make an elegant twist, Spanish-oriental style, cup your hand and gently pick an orange. Lower your hand again in front of your face, wrist first, down to belly height, open your hand gently as if dropping the orange in the basket, hand back on hip. At the same time the other elbow initiates the arm and hand movement upward, wrist and hand twist, cup and pick, go down to the basket and back to the hip.

When both hands rest on the hips again, the first hand and arm do a similar movement starting from the elbow, now reaching forward. The same with the other arm and hand. Then the first arm and hand do a similar movement out to the sides, then the other hand and arm. The movement always starts from the elbow and the wrist, soft and with bound force, as if the joints move in silky honey. The eyes follow the movement, glancing up, forward, to the sides.

Repeat the movements at a swifter pace, up, forward, to the sides. With the one hand, with the other hand, rhythmically: *up* and *up* and *front* and *front*, to the *side*, to the *side*. The quality of movement is as soft and graceful as it is firm and articulated.

Interlude with handclaps and leg kicks.

Do the basket again, the figure of eight around the hips and behind the back. Then lift both hands along the sides of your body all the way up, stretch the arms and your whole body up, rise on your toes, stretch a little more, hold … hold … Fall down, curve your back, bend your knees, head down, hands on the floor beside your toes. Support with hands and feet and sway your backside gently up and down, to the beat of the music.

Crawl forward on your hands, toes staying on the floor: one two three four five six seven … stretch your body like a board, press and lift your hands from the floor, clap on the eight and shout ‘Hi!’, only the toes keep floor contact, hands quickly back on the floor, stay in the rhythm. Crawl back on your hands, one two three four five six, back on your heels on the seven, clap your hands on the eight. Repeat the walk on the hands forward, do the board, clap on the eight, ‘Hi!’ Go back: one two three four five six, back on your heels on seven, get ready, arms flinging backwards in big arches on the eight, hands land on the floor under your body, back turned to the floor, belly in the air.

Walk into the ring on all four: one two three four five six seven … stretch one leg out, kick out on the eight, flex your foot, shout ‘Hi!’, crawl back: one two three four five six seven … sit down on the eight, legs crossed.

Picking bananas with a snatch; up, front, to the sides.

Do the basket on the floor and the circle around the hips, a lying eight, hands crossing front and behind the back. Again, do the basket, hands cross far forward, back in front of the belly, back to the hips.

As above: lift one arm, elbow first, then wrist, hand and fingers hanging down, going up in front of the face, unfold the arm until stretched. High above the head the wrist and hand make a Spanish-oriental twist, the hand cups, the fist clenches. Hold a second, drag down a banana with a little snatch—put some energy in your voice, as if you were sour and surly: ‘A …banana!’ Bring your clenched fist down in front of your face, down in front of your belly, drop the banana into the basket, hand continues to the hip, the other hand starts the movement up, picks with a snatch, goes down over the basket and back to the hip.

Repeat the movements, now forward with a stretched arm and a clenched fist, the little snatch, back again. One hand, the other hand. The same to the sides. The movements are initiated by the elbow and the wrist, soft but distinct, with a little accent in the snatch, everybody acts sour and surly.

Repeat the whole sequence at double tempo, rhythmise: *up, up, front, front, side, side.*

Picking small berries and berry fighting.

Do a tiny little basket on the floor, with your forefingers, elbows stuck to the body, indicate the arch around the hips and behind the back with your forefingers. Pick small berries, tiny, tiny small berries, pick them with your fingers, by the dozen, everywhere, close, a little further, up, forward, to the sides, next to your body. Pick, put in the basket, another berry, one more, one more.

Glance sideways at the one sitting next to you, lean over and pick a berry next to them, put it quickly into your mouth, hide your hands behind your back quick as lightning, press your lips together, sit dead still for a second, staring straight in front of you. Glance sideways at the other side, pick a berry next to your neighbour sitting there, put it into your mouth, hide your hands behind your back, mouth shut, gaze still. Pick up a berry, throw it at someone in the ring, freeze your arm in the air and spread your fingers half a second. Pick up another berry and toss it at someone else, another one, and another one, throw at everyone, all over, fling out your arms, throw, throw, throw …

Wiping our faces.

Freeze your body, face and arms in the middle of the last throw. Screw up your eyes, make faces. Put your hands in front of your eyes as for protection, elbows splayed out. Do a wiping movement with both hands from your eyes and out to the sides, as if wiping off smashed berries with your fingertips. Shake your hands and fingertips over the basket, wipe more berries out of your eyes, shake off your hands down in the basket. Shake your hands and fingers with bigger and bigger movements, arms and shoulders shake along, the whole body shakes. Rhythmise, like before when we shook off water drops from the tips of our fingers: shake, shake, shake, hold! … Shake, shake, shake, hold! …

Hold your hands in front of your face, palms turned to your eyes, pinkies together, elbows out. Do a wiping movement: down, up, down, up, while at the same time the face goes up, down, up, down. Move your hands in front of your face from side to side, the face goes in the opposite direction. Rhythmise on a 4-beat: one two three *freeze*—hold on four, look in the opposite direction—one two three *freeze*—hold, look in the opposite direction—one two three *freeze* … one two three *freeze* … Keep your hands in front of your face, make small wiping circular movements in front of the tip of your nose, elbows sticking out, bigger and bigger circles, bigger and bigger, still bigger, as big as you can, the whole upper body swaying and tilting in all directions.

Picking watermelons and getting a thump in the head.

Talk with a noisy, drawling voice. Do a biiig basket, do a biiig riiing around your waaist, haaands behind your baaack, haaands up in the air, piiick a waaatermelon high up in the air, it's veeery heavyyy, liiift it dooown into the baaasket, the whole body craaashes down. Pick another waaatermelon, high up over your head, down into the basket. One mooore waaatermelon, this one thumps down on our head—'Ouch!'—down into the basket. One last one, thumping down on our head—'Ouch!'—down in the baaasket.

Rising up from the floor and stamping our feet.

Do the figure of eight, the basket and the circle around the hips. Once more, the basket, the ring around the hips, hands behind your back. Now the hands sweep around and under the basket, lift the basket, put it on your head, it's very heavy. Take away your hands, hold them together palm to palm at chest height, balance the basket on your head, try to get up, balance, balance, keep your head still.

Straight backs and necks, snap your fingers, lift your eyes to the basket, spin your body, round, round one turn on the spot, carefully. Lift the basket from your head, throw it up in the air with all the oranges, bananas, berries, watermelons. Hands on hips, stamp, mash.

Dance Like Another

After an intensive, style-bound and instructed dance exercise like the one above, with recurring sequences with different variations and qualities of movement, wrap it all up, transfer to free dancing and leave the initiative to the dancers. Run through all body parts, from the fingers dancing to the whole body, smooth and escalating. Let everyone dance like someone in the group, then like another, and on through the whole group.

Let your fingers dance freely, then fingers and hands, then fingers, hands and wrists. Elbows join in and dance, shoulders, chest, neck, head, face, hair, hips, thighs, knees, lower legs, feet, toes, the whole body dances. Dance freely. Dance like Julia. Dance like Kaspar. Like Lilly, like Doris!

Roundup and Relaxation

Rocking and Rolling Waves

Music: sea rush, waves, e.g. Radio Tarifa: Soledad; helping tools: big fan.

Everybody lies on their back, spread all over the floor.

Close your eyes. Put your hand on your heart. Feel its beat. Our hearts make it possible for us to move, pumping blood to all the muscles in the body so they can grow and get stronger. Meanwhile we help keep our muscles supple and smooth by dancing. Think of everything we've done and danced today while our hearts beat thousands of beats. Remember how we galloped, turned, hopped, fell down. Feel the warmth in your body, we raised the heat with every movement. Sense the floor against the back of your head, against your shoulders, your spine, the curve of your back, your sacrum, your behind, thighs, under the hollow of your knees, against your calves, your heels. The floor that bears us. Close your eyes. Let your whole body's weight sink down on the floor and to Mother Earth.

Go around with a big fan, fan over all the dancers lying down on the floor.

Make your body light. Light like a feather. Like when we flew and let the air carry us. Feel the breath of air over your body, over your face, forehead, eyelids, nostrils, lips, neck, breast, stomach. Over your legs, knees, wrists, feet, between your toes. Make your body light. Light like a leaf in the playground. Let the fan blow the leaf away a bit.

Make the body light. Light like water drops. Let the wind gently rock you from one side to the other. Roll over when the fan comes sweeping by, roll over like small waves on the sea. Roll back, and roll over again. The wind rises, let it roll you over all the way, roll over like a wave washing over the beach. Your body, arms and legs, your hair wash over the sand. Roll back, the wave pulls back into the sea. The wind takes off. Let it rock you softly, less and less. Let it rock your dance body gently.

Relax, close your eyes, lie and rest, lie still on your back, breathe.

Read Strindberg (1902), from *A Dream Play*:

We, we, waves,
are rocking the winds
to rest!

Green cradles, we waves.
Wet we are, and salty;
like flames of fire;
wet flames we are.

Dousing, burning,
washing, bathing,
breeding, bearing.

We, we, waves,
are rocking the winds
to rest!

Open your eyes. Sit up. Thank you for all that dance.

Let's Talk Dance: Meta-Talks

A *meta-talk*, in which the participants look back and talk about the activity done, is a pedagogical tool aiming at raising awareness to a cognitive level, reinforcing insights, knowledge and skills which they gave proof of and deepened into during the activities.

Meta-talks in dance teaching and learning enhance greater kinaesthetic awareness and kinaesthetic literacy. As the young dancers find words to describe what they did, using appropriate dance language, they catch sight of their aesthetic abilities and their ethical responsiveness in the interplay, finding inspiration and care for the joint creative activity.

Acquiring a language for dance is in itself general language and literacy developing. Mastering dance's specific vocabulary helps moreover to look at things through the dancer's gaze. When we visit an art event or exhibition with a preschool group or a class, we open for aesthetic experiences. Our senses work through the artist's way of composing a painting or a sculpture and how these are hanged or placed on exhibit for us to experience spatially. We can ask the children if there are figures to discern. Are there different fields of colour? Soft or raw surfaces in the stone or wood? Are straight or curved lines used or none at all? How does it feel to stand in front of a picture covering a whole wall or before one as small as a postcard or a stamp?

In a similar way we can take a closer look at how a dance is done, as a whole as well as into the smallest wink or little thumb's movement, how the dance starts and how it ends. Talking together and naming aspects of dance is a way of getting acquainted with the language of dance.

During a meta-talk after a dance exercise the young dancers, finding the words to say what they explored, gain additional access to dance as a specific mode of knowledge and insight, a mode of art and a way of being in the world. Dance language is the language of the dance art form where the words help us to see new perspectives and new possibilities to move towards. Words naming body, space, time, force and interplay wander from everyday contexts into the world of dance and are given new flavour and meaning. The words and idioms help us communicate the power of dance and to see the world and our existence in it in a new light.

When conducting a talk about a dance activity, go into particular movements and sequences, the feeling in the limbs, images that arise, the way others moved or looked like. Help find the appropriate words for the physical, the kinaesthetic and the transcendental aspects, the aesthetic and the ethical.

> Let's have a chat about the dance. Let's put words to everything we did.

Ask open questions:

> How did it feel in your bodies? Is there anything of this you have done before? What was new? Which movements were new? Did it take an effort? Was it fun? Did you get wheezy? What would you like to do again? What would you like to do more of? Did you see someone do something in a whole new way that you would like to try out? What do you like about dancing? What is dance really?

Ask directed questions:

> How did it feel to do skipping steps backwards? What is a twist, can you show us? What does bending and stretching mean? Which muscles worked the most when we flew around in the quiet dance?
>
> What went through your minds when we galloped and jumped over the fabric? How does it feel to dance with your backs against the wall? With your backs turned towards the middle of the floor? Backs turned to each other? To dance in the centre of the room? Do you remember where in the room we were when we did the basket? When we galloped towards each other? When we walked with our whole bodies? Which one of the movements was the fastest of them all? Can you do such a fast movement really, really slow? What would be the smallest movement one can do? The biggest? When did you get a chance to show how strong you are? Which muscles got to work the hardest when we did a hop? When we squatted down and then rose up again? How do you picture a movement in your mind?
>
> How do you go about putting together a dance? Is it easier to do it together with someone? Or is it better to just be left alone and come up with moves by yourselves? What is a solo dance? Does anyone know the word 'solo' from another language?

Doing your Own Thing

In terms of teaching mode there is a good degree of demonstration, instruction or *direct mediation* of movement in the dance activities described above. There is room for the dancers to fill out with their own interpretations, ideas and impulses. Teaching dance at large, as mentioned before, aims at the young dancer's capability and proficiency in creating dance themselves.

Encourage even the youngest dancers early to put together movements or sequences and make choreographies, as they acquaint themselves with a repertoire of basic steps and explore these in a thousand variations. Give them opportunities, through improvisational and compositional work, to try out kinaesthetically appealing movements. These may be moves they may experience as 'cool', 'nice', 'cheeky' or 'awesome', and may want to beautify or uglify more—'ugly dancing' is a tremendous source of joy and bold ability to put together a dance.

Throughout, the aim is to increase the young dancer's ability to play with dynamics and contrasts, such as between forceful and holding back, large and small, fast and slow, rhythmic and free-flowing movement, between dancing solo and in pairs or all together, either synchronised or moving differently. There are numerous exercises to get started at compositional work in dance.

After this section on teaching creative dance, the next chapter deals with dance teaching entirely oriented towards 'doing your own thing'. 'Taking after' or imitating will continued be affirmed in the sense of trying out others' movement material. Some of the exercises described above will reoccur in a modified fashion.

Chapter 3
Dancing with Each Other

P. Moerman, *A K-12 Dance Method to Teach Creative Dance, English, Math and Science*, https://doi.org/10.1007/978-3-031-93156-7_3

Dancing with each Other is a dance teaching method with a double artistic and educational aim, intimately intertwined. Creative dance is taught hands-on through improvisation and composition. The emphasis is on the joint venture: to explore and experience dance and its power as art mode together, in the preschool group or in class, and to take on a shared ethical responsibility to provide space for everyone to come forward and bring their aesthetic input into the dance work.

The ambition is to introduce and practice dance from its basics and in its richness of movement variation, and to allow for encounters dancing and interplaying. The dance teacher does not demonstrate any steps and presents no themes. The point of departure is mere movement, coming up in the improvisation and compositional work in whole groups, in smaller groups, in pairs and one by one. Eventually everyone creates a solo which they share and teach to each other as a foundation for a joint choreography.

> The undetermined, the unpredictable and the unintentional in 'doing' dance are prioritised. The aim is not result, performance nor entertainment, not thinking 'outward' towards a beholder or judging gaze, but searching 'inward', indulging in movement and the inner and interhuman experience of dancing and creating together. Means and end are the work of the senses, a bond of receptivity and activity, presence and concentration.

Dancing with each Other is conceived out of discussions with young children about dance play. Often they express clearly existential reflections, in simple words such as: 'I am my dance', 'When I dance, I am like someone else', 'In dance you can do what you want', 'I dance when I am happy' or 'I'm happy again when I dance'.

The teaching method aims at opening up for what 'pure dance' can show and 'teach' us. What is it like to dance with others. What is it like to dance and bring the power of imagination and action into play. What is it like to take the aesthetic and ethical responsibility to welcome each other, each and one in their unique, dancing and human otherness.

The method invites to dance for the sake of dance itself: to scale down dance and lay bare its quintessence, beyond representation, interpretation and even expression. In search for 'dance as dance' and the pure kinaesthetic sensation of movement, beyond intention and cognition, ultimately there is only the dancing person, dancing and appearing in front of and with others—the core of dance art. The participants catch sight of each other anew—and of themselves.

Together the dancers create a world of movement where everything can be different and anything can happen, right in the middle of everyday school life. They create a dance universe, a tangible world of possibilities, an indispensable alternative to an order with little room for manoeuvre otherwise, where all is as it is.

This way dance creates a space where each one can bring something of their own to their education and to the world, something unique, created here and now, something that cannot be prescribed in a curriculum, something that comes out of each and one's true freedom to act. *Dancing with each Other*, they create intersubjective spaces of becoming: spaces for the relational and existential dimensions of dance—and education..

Improvisation

The empty floor as metaphor and 'for real' marks the trend from the outset—a place where every move is a new beginning. A range of exercises are suitable to playfully start filling the floor with movement.

Say Hi with a Movement

Everyone presents themselves in turn with their name and a movement they create on the spot.

Start standing in a circle or spread out over the floor. Ask for a starting position, a simple form, open and neutral, relaxed and stable:

> Stand upright with parallel feet. Or stand with your weight on one leg and your hands on your back.
>
> Everyone in turn says hi and their name: 'Hi, I am Molly!' and makes a movement of their own choice. All respond: 'Hi Molly!' doing the same movement. The one who said hi says 'Thank you!' and 'There you are' to the next one, with both an inviting gaze and gesture. Everybody resumes the starting pose. The next one says hi and their name and makes a movement of their own, all answer saying hi and the name and doing the movement. Encourage the dancers to go along with the first best idea, a curtsy, a wave, a stamp, a turn, two claps in the hands or on the floor.

The return of each greeting is a response, reinforced by the repetition of the movement. The tone of voice and detail of gesture echo back like a unanimous *yes!* A welcoming space opens for everyone's initiatives, a space of reciprocity.

Dance Relay

Another way to start, with the same message about the value of everyone's input, is through a relay of movements, receiving and passing on initiatives.

It is also a simple way of choreographing a dance together where everyone contributes with a move.

> Everybody stands in a circle.
>
> Someone shows a short movement sequence with a clear beginning and a clear end. Repeat if needed. All do the movement jointly, freezing the end position. From there the next in the circle is ready to add a movement of their choice. Repeat the first movement all together, hold the end pose. The next one does their movement and freezes its end position. All do the first and the second movement. The third one adds a movement, freezes the end. From the top again, first, second, third movement, hold, number four adds a move. Each end position is held so the next one's onset is clear to everyone. Always encourage the dancers to go for the first best idea. Repeat from the top after every new contribution. If the group is numerous

you can repeat just three or four moves back in the order, let new dancers add their movements, and then take it all from the beginning now and then.

Rehearsing the entire series of movements from the beginning as the relay expands works surprisingly smoothly also in a bigger group. In each link of the chain all eyes are on the original contributor, a photographic aid to the muscle memory.

Add a rhythm to the first contribution and establish, for example, a simple common time. Be open to sounds, air puffs or smacking noises, hand claps or feet stamps, which all used to come up in the playful creative work.

When the last dancer has made their contribution and the whole relay sequence of movements has been repeated a couple of times, put on the music, rhythmical and melodical or a soft and airy tune, and do the whole dance a few times.

Then encourage a round with greater force and vigour. Go for it all the way, with a sharper rhythm or staccato, or slow-motion. Then try and do the whole choreography backwards. It usually works out fine and creates a feeling of being able to cope with more dance than one could imagine.

This exercise in itself is one of many ways to create an entire and unique choreography building on every participant's single contribution of mere movement.

Follow John / Yoanna Taking Quick Turns

Music: free choice, calm or driving, pop, world music, e.g. Warda: Batwanas Beek.

A warm-up can also be achieved with speed. Again, the attitude is mediated from the start: cooperation and joint aesthetic and ethical responsibility, for the dancing and for each other.

All stand in a file behind one another. The one standing first dances in whatever way and wherever in the room they wish. All follow along and repeat as carefully as possible, observant to details, absorbing the flow of movements as they are invented.

Everybody takes responsibility for the one dancing up front, by relieving them after a while and taking over the initiative. One after another takes the lead, whenever they feel ready.

Each new leader gives the dance a personal touch, spurred on by the joy of invention and the collective response.

Someone swings their arms energetically. Someone else makes a sudden stop. Another sits down or lies down and continues dancing with their legs or toes only. Someone makes the whole file move backwards. Someone jumps up on a bench and dances as long as it is and jumps down again, or dances out through the door and back in again. Still someone else claps their hands, stamps their feet or makes sounds with their voice. Movements come up coloured by hip-hop, oriental dance, ballet, folklore dans, circle dance, even sports and gymnastics movements. Everybody follows the flow, every movement is born in the instant and spreads instantaneously through the winding line.

When everyone who wanted to has danced up front and the ice is broken, turn the music off and sit down. Take a breather and talk about the dance.

Was this dance? What is dance? Where did all that dance come from? From us!

Draw attention to the richness and abundance of movement coming out of the group. A medley of ideas, details, genres and temperaments filled the room. Point out the rapid communication: at the blink of an eye every movement is transmitted along the line of dancers. Movement and rhythm are reproduced practically in the same moment they are perceived, like in an attuned orchestra. Kinaesthetic perception evolves instantaneously between one's own body and other bodies in movement.

Also this warm-up exercise provides plenty of movement material to compose a whole dance.

Lead, Follow or Swap

Not everyone feels comfortable taking the lead and having a line of dancers at their heels. Make it clear that no one has to. Dance, in all its wonder and potential, can hardly be forced upon any child or student. That would entail taking away the individual's subject-ness and freedom, central aspects in the overall argument of this book and everything being strived for in these exercises. Hesitation, insecurity and resistance must be met educationally, in every sense: resistance and risk-taking are preconditions for transgression in meeting the unfamiliar and the unknown. First and foremost, doubt and hesitation must be addressed with adequate teaching method tools.

Dancing in front of others means taking a risk. Daring to take a risk draws on trust. Wait for those who are reluctant and establish a trusting atmosphere and conditions that encourage a willingness to take on risk—the basis of all learning and growth. Tone down expectations of performance in favour of play and trial.

One subtle way to facilitate initiative is to divide the group into three.

Everyone gets a number: 1, 2 or 3, indicating one's range of capability in the exercise. The ones can only lead. The twos can only follow. The threes can switch: at times they can choose to lead, then shift and follow someone.

Put on the music.

Everyone dances according to their number and ability. Each one needs only to remember what they themselves can do or not do. The ones move independently from the others. The twos hook up with anyone who comes dancing by. The threes shift between inventing and moving along.

The exercise is another example of how to create rich choreographies with a few simple instructions that facilitate the alternation between taking initiative and moving along. Many choreographic moments arise from pure movement and movements' power to lead to unexpected patterns and occurrences on the floor:

A flux ensues in the room as groupings are shaped, lines and figures are formed and accelerations occur. Flows constantly shift and dissolve as new ones arise. At one point someone is dancing in stillness all alone in a corner, while a larger group has formed a circle in the middle and transfers into a pulsing row billowing through the room. A cluster of four is standing at one spot, dancing with one body part at the time, someone disengages and gallops away, gets a follower, they speed up across the floor with huge springs and leaps. The dancer in the corner has merged into a new, hectic and stirring constellation.

The choreographies shift incessantly like a swarm of starlings in the sky or patterns in a revolving kaleidoscope.

In the continuously changing dynamics in the dancing, those who would rather wait and see might find opportunities to take initiative and try out their own movements, without being in the centre of attention.

Dance Relay in a Circle

Music: rhytmical, e.g. Monsieur Blumenberg: Requiem pour un melodie oublieux.

Everybody is standing in a circle.

Someone goes inside the circle and dances to the music, in their own way. Everybody in the circle dances like the one in the middle. Soon someone joins in and dances with the one in the ring. When they have danced together a while, the first one backs out while the second one continues in their own way. All the time everyone in the ring dances like the one or those in the middle. A third one steps in and follows the second one's moves. They dance together for a while before the second dancer backs out. Then the third continues alone and a fourth comes in, dances like the third, continues alone, and so forth.

If two people happen to step in at the same time all three dance for a while until the first one steps back. If someone is left alone without anyone joining in, the lone dancer can dance towards somebody in the circle and invite them to come along—often they do a rope-pulling gesture or haul in an imaginary casting rod. If the lone dancer does not get a bite they move over to a new possible partner. Or you join in.

When all or most of the dancers have been inside the circle taking turns, ask everybody to join in and dance like the one or those dancing in the middle, who eventually find a way to end the dance.

Talks About Basic Steps and Building Blocks of Dance

Revise basic steps together. Ask for an example.

Skip. Ask the one who came up with 'skip' to show. Ask everybody to skip in the same fashion. Ask for ways to vary the skipping step, with respect to body, space, time and force: skip on the spot with high knees; skip backwards with hands on the sides or hands behind the back; skip with high arms moving up and down alternatively; skip with delicate leg movements or the whole body involved; skip backwards, forward, round and round, and so forth. Ask for more examples. Walk, jump, turn, sneak. Assist in seeking variation and contrast – different movement qualities. Ask everybody to try a movement or a step sneaking as slowly and silently as possible, making the body as small as possible, or filling out all areas and levels of the room with the whole body in gigantic movements.

Reaffirm that each step and each movement we come across can be danced in an endless number of ways by thinking body, space, time, force and relationships between these basic elements and the others dancing.

Improvisation to Music

Music: varying, complex in pulse and tone, e.g. Joaquín Cortés: Desafian.

Half the group is sitting in a big spacious circle with ample room between each person for the other half, spread over the floor, to move all around. The half not sitting in the circle is sitting or lying down, closing their eyes.

> When the music fills the room, let it into your ears. Listen closely to how you feel it wants you to move. Start to gently move your body that way. Open your eyes slowly, rise and go on dancing as if the music whispers in your ear. If you sense it wants you to dance with fast feet, try it. If it wants you to dance as quietly and slowly as you can, go for it. Try a big leap or small hops if it sounds like that is what the music wants. Fall down, roll over the floor, fence with your arms or spin around inside the circle if you just feel you want to try that out.

Turn off the music. The dancers go and sit down in the circle while the other half spreads out and gets ready. Put the music on, the second half dances. Afterwards discuss how the dance and the music can go hand in hand with each other.

In the next round the half groups get to dance the other way around.

> This time don't 'obey' the music, but see how you can 'defy' it's rhythms and feelings, offer resistance, or move in dialogue with the music.

Mirror Exercises: Two by Two, Four by Four, Eight by Eight, All Together

A classic exercise two by two unfolds into a movement dialogue. In this version the exercise evolves as the couples eventually merge into quartets and these in turn merge into octets and so on. In the couples one initially leads while the other mirrors the leader. Then they swap roles so the one who first 'spoke' now 'listens'. Then the interplay transfers into a free dialogue of seamless 'speaking' and 'listening'.

In all dance improvisation a readiness for the unexpected is practiced, coming from one's own imagination as much as from other's ideas. Encourage attentiveness, docility, play and interplay, as well as the collaborative richness of invention.

> Everybody is standing in pairs, all over the floor, two by two facing each other. One in each pair starts as the leader, the other one mirrors the movements. The leader takes responsibility so the other one can follow along, starting soft and easy with small simple movements on the spot. As the interplay is established, the leader elaborates into a larger scope of movement variations and qualities, moving closely or further apart from each other, in stillness or with intensity, and so forth.

After a while the two in each pair switch roles. After another while the interplay evolves into a keen communication where both speak and respond, initiate movement and follow the other's impulses, give and take.

When the pairs have danced for a bit, filling the room with motion, they merge two by two into quartets. The steering and mirroring continues seamlessly so you can't notice who is sending out or receiving an idea. After some time the quartets melt into octets and continue the interaction eight by eight.

Let it take time. Encourage eye contact, silence and concentration, and a slowing down when it becomes overly elaborate and too tricky to hang on to. Urge those who are eagerly feeding in their own ideas to go with the flow and follow for a while.

Finally all dancers synchronise the movements born in the instant. Ask them to orientate the interplay towards a conclusion, or little by little decrease the movements down to the tiniest possible and then stillness.

Contact Improvisation

As the young dancers get a hold of their dialogical capacities doing mirror dances, they can engage in exercises with different degrees of physical contact.

In contact improvisation two or several dancers maintain body contact with each other without holding hands. What is explored is the interplay between one's own weight and force and the other's. As in the mirror exercises the interaction draws on sensitivity, attentiveness and flexibility, the dancers now moving in greater proximity and taking on bigger responsibility for each other.

In the simplest paired exercise one dancer places a hand on the other's back, between the shoulder blades and with firm contact from the finger tips down to the heel of the hand. Through mild pressure on the peer's back the leader steers them through the room. At first suggest a plain walk to get acquainted with the dynamic. When the hand presses on in a more determined fashion the movement shifts forward with greater force. When there is only a minimum of contact without the slightest pressure, both stall. When the hand slowly retreats, the dancer's back also retreats to maintain the communication of touch.

Gradually the one who is steered senses the smallest twist of the other one's hand and turns into the intended direction. If the hand presses downward, the steered one goes down by bending the knees—so does the steering one, echoing each move. Then the downward pressure continues all the way until sitting. Throughout the whole process the steering partner mirrors the steered partner's movements and positions, as in tandem.

Sitting back-to-back is another simple exercise in pairs. Both feel each other's weight, first just bending and straightening their backs, then leaning to the sides and in circular movements. Then they try to lift their buttocks by keeping their feet steady on the floor and pressing their backs straight and firmly against each other, rising together and dancing on through the room back to back.

Still another paired exercise is moving over the floor maintaining contact only between the shoulders, the shoulder blades, the hips, one side of the torso or one side of the head.

Composition and Choreography

Keep reminding the young dancers of how they can perform every step and movement in endless variations according to how much body they engage to shape forms or where in the room they let the dance evolve, how they can play with speed and rhythm, how much muscle force they deploy or how they can mimic or contrast other's ways of moving.

Thinking this way is thinking like dancers and choreographers do when putting together a dance piece. Composing in dance or choreographing can be done in many ways. The improvisation work described above are in themselves ways to invent ingredients in a choreography. There are many simple ways to go about composing.

Composition of a Four-Step Dance

Help the young dancers name all steps and use appropriate dance language to describe the various ways in which the movements can be danced.

Ask for an example of a basic step.

Gallop. Ask the one who made the suggesting to show the gallop step. Let everybody do it that way all together. Ask if the same step can be done in more ways. With high knees. With almost straight legs. Straight across the floor. On the spot. Backwards. Hands to the sides or in front of the body like a driver or a jockey. Let everybody try out all suggestions as they come up.

Ask for other examples of elementary steps and explore variations.

Ask the children to imagine a dance made up of four parts. Each part is a movement performed *bodily* in a certain way, using *space* in some way, with a certain *speed* or *rhythm* and given a certain *force*. Ask for four examples of such parts or demonstrate one or more yourself. Ask for descriptions of the suggestions:

What is this dancer's starting position: where in the room will they start that part of the dance? Which direction are they facing? Where do you think they will take off? Are they relaxed or are the muscles tensed?

Standing in a corner. All straight upward. Feet together. Hands behind the back. Looking straight ahead. Towards the middle of the room. Towards the opposite corner.

The first movement.

> Gallop forward. Hands kept behind the back. Lifting the knees a little. Halt in the middle of the room. Was there a leg change? Or was the same leg in front the whole time?

The second movement.

> Spin around. On the spot, in the middle of the room. Arms stretched out sideways. Not fast, not slow, in between. Carefully. All silent. How was the face, the head? Smiling. Staring straight ahead. Looking at us, then spinning and looking at us again after every turn around. The head went around fast and the eyes looked over here again. How many turns were there? How did this part end? Standing still. Like a T. Facing us watching.

The third movement.

> Jogging. Round, round, on the spot. High knees. Arms moving along up and down. Elbows too. Hair too. Clenched fists. In which direction did these turns go? The opposite direction from the spins in the second movement. How many turns this time? How did it end? Straight up and down. Side turned towards us. Arms hanging besides the body.

The fourth movement.

> Seal hops! (Hopping up, throwing legs backwards, bums upward, chest down, body falling flat like a board, landing with the hands on the floor, then chest and stomach landing, then toes touching the floor.) Lying on the stomach stretching out arms and legs like a stick, pointed feet. Rolling away to the wall, against the congas. Lying still.

Ask the dancer to dance the whole composition with the four parts one after the other. Ask everybody to watch closely again and check the different movements and positions. Then all dance the same little dance, in half groups. Put on the music. Clap the beat together with the watching half.

> Divide the group, four by four. Each quartet creates a dance with four parts. Everyone contributes a movement and shows the other three how and where in the room it is done. The four parts are put together in succession. Make sure each movement is given time to take place as a dance part. Do not rush and scramble through the movements, let everyone get a kinaesthetic sensation of each contribution.

Play some music in the background and let the groups work in peace and concentration. Go around and spur them on, reaffirm the ideas that come up, comment on qualities of movement, point out possibilities of variation. Encourage them to fling out a gesture, or to hold back.

Suggest forms for the quartets to put their four parts together. Either they all dance the four parts one after the other synchronically, or two by two dance two of the movements simultaneously. Maybe the four dance in a row, in a circle, in a square or however they may choose to solve it.

As they acquaint themselves with this way of working and composing simple choreographies, the task can be diversified. One group looks for spatial contrasts: making various shapes with the body alternating with keeping the body straight; shifting between sticking to one spot and spreading all over the floor, reuniting at one point and then dancing helter-skelter; dancing two by two moving away from each other and meeting again; one stays in the middle while the others orbit around, and so forth.

Another group varies between very fast and very slow, between repetition and stillness, between different rhythms, including clapping hands and tapping on chest, shoulders and hips, and so on. A third group explores contrasts between intensive movements so the whole body shivers and more restrained movements, between robotic and zombie-like movements. A fourth group investigates dancing with individual body parts—hands, hips, torso, feet, face—in contrast to engaging the whole body, vividly or in a suspended mode.

As composing skills and an awareness of the elements of movement increase, the tasks may be further refined towards investigating combinations of two parameters; for example, body and force or time and space. They might start from a creeping movement and accentuate its flow through a sudden lunge, or drag it out as long as possible, inch by inch, at the pace of a snail about to fall asleep—play and explore.

The relationship aspect is present throughout this dance exploration work. Firstly on a mere movement level as the parameters space, time, body and force always naturally intertwine. Secondly in the building of relationships between the dancers. They affect an openness and willingness to be inspired by each other's suggestions and to try other ways to colour movement. Or they explore their own interpretations sampled on each other's steps. All this nurtures an overall atmosphere of jointly creating movement, meaning and communication.

Meta-Talks on Choreographies

Watching each other's compositions together offers great opportunities for inspiration from each other's ways of creating dances. No dance teacher or choreographer could come up with such a rich range of samples and possibilities just handling the building blocks of dance. Discussion deepens the young choreographer's awareness of dance, movement and composition.

Direct attention to all aspects of the dance and to finding words to describe its features. Avoid using judgemental terms when discussing the dance work. Direct the conversation at *seeing* and *experiencing* dance, at *how* each group enacted a task. Ask questions:

> How did they start their dance? Where in the room did they move? How did they displace themselves? Did anything happen suddenly and surprisingly as we were watching? Which body parts were busy the most? How did their dance end? Did they seem calm? Were they stony-faced of bubbling with joy? Did you get to think of something special when you watched them dance? In what kind of world were they?

Dancing a Solo

Everything touched upon so far, about composition and choreography, about dance and subject-ness, both from the creator's and the beholder's point of view, comes to a head when only one person dances.

This can only happen if the dance teacher and the young dancers succeed in creating a trusting environment, so those who are hesitant and resistant dare take the risk to emerge with their solo.

The task to dance a solo may feel insurmountable. One step on the way is to choreograph in smaller groups, like in the previous exercise where each dancer contributes with their 'part'. The solo task also becomes more doable when the dancer can start from something simple and concrete, like the characters in one's name. Another rewarding idea is to start from those small habitual tics one uses in everyday life:

> Pass a hand through your hair. Rub your nose. Wring your hands. Drum your fingers. Adjust your shirt, collar or waistband. Shake your knees. Shrug your shoulders.

Other suggestions are to dance a colour, or one of the basic elements earth, fire, water and air.

The purpose is not illustration or documentation. It is all about searching for freedom and imagination of movement. It is about free association and interpretation of these curlicues, letting them take unexpected directions and seek variation in body and space, time and force.

In a solo starting from the letters in one's name the graphic aspect is not primary. It is not a matter of 'writing' or 'drawing' the characters in order for the spectators to be able to 'read' the dancer's name. What the letters offer is precisely—and literally—a graphic *form* to play with and explore possible *dancing* forms through.

> Allow imagination of movement, follow impulses, vary degrees of speed and intensity, spatiality and use of the body, insert repetitions, stops, slowdowns and accelerations, do sequences backwards, in another order.
>
> Someone does do the O like a tiny little hole in the fist to spy through, quickly squatting down; or dances a blown up O, the whole body bouncing around the room like a big airy ball; or dances the O like a giant evening sun slowly setting. Someone dances and puts the dot on the i or under the Arabic ب (*ba*) in different spots here and there, minimalistic through a blink of the eye or the entire body's weight crashing on the floor. Someone does a U skipping over the floor arms up in the air. An L stands leaning the body against a wall, arms up along the wall and only the toes move up and down, or the balls of the feet tap on the floor.
>
> Let each movement give new inspiration to take the dance further. Try to do a movement squiggle three or thirteen times in a row. If you are drifting further and further away from how an i or a ب or a U or an L look like on paper, then you are well on the way, just keep on dancing to your own ingenious pipe.

Exploring movement in this way, starting from letters, shifts the focus from 'outward' to 'inward'. From worrying whether those watching will succeed in deciphering the letters in your name, to caring for one's own kinaesthetic sensation doing the movements. In turn, and only then, it caters for a reciprocal kinaesthetic experience in the beholder. The young dancers, instead of reducing themselves to objects for others' examination and judgment, take place as subjects capable of acting and accomplishing their own movements, their own particularity, their own subject-ness.

Further down all these singularities gather in one choreographic work.

Let everybody work separately spread over the floor. Ask them to leave each other in peace and promote silence. Have some soft music in the background. Encourage the dancers, mirror someone's movement, freeze it and ask for suggestions where to go on from there.

> Let them carry on with the work, until inspiration runs out or they seem to have put together their dance. Then everyone finds a spot on the floor and takes a position suitable to start from. Let them close their eyes and think their dance through from the beginning to the end. On a signal they begin together, all concentrating on their solos. When someone is done they keep standing, sitting or lying still until everybody has finished their dance. On a new signal all dance their solos again, this time very fast, like a fast-forward film.

Then everybody sits down, either just where they are, spread over the floor, or along the walls leaving the floor empty. Time now to see what everybody has been working on. Clarify once again that this is not a performance or a review, but yet another step on the journey into the world of dance. It is a big step, creating a piece of dance all by oneself. So now let us just see how it has gone after only a little while of playing and putting together a solo.

What we are engaged in is exploring movement. Let us look at the floor as a dance lab. Cherish the moment when taking the floor, allow space for further inspiration and improvisation, just be there, in your dance. Likewise, cherish the moment when watching a co-dancer fully present in their dance.

For that reason it is a good thing to refrain from applause after each solo. Applause reinforces the feeling of performance and the dependency on approval of a result from without. It risks breaking the intimate atmosphere of free trial created for one another. The focal point is the sensuous experience of movement of both doing and watching as one after another gets up and dances—the kinaesthetic sensation, the *magic* of dance.

Save all emotion until the end when a resounding applause acknowledges the important brave work that has taken place in front of our eyes, witnessed by all.

Make it clear that no one is forced to dance their solo. Of course it may feel scary to get up and dance all by oneself in front of everybody. More important and exciting is to see how different everyone's solos are and that nothing can be 'wrong', never ever.

With older children and students you can articulate the nature of this essentially educational, artistic, relational and existential moment, as mentioned earlier in the theory section. This moment deals with the decisive question of risk-taking, a moment present in every pedagogical context as a prerequisite for growth, insight and becoming. Ask those facing the dilemma to take the risk or to refrain, to consider these two questions:

If I take the risk to dance my solo, what can I gain from it, and what can I loose from it? On the other hand: If I refrain from taking the risk, what can I gain from that, and what can I loose from that?

Then let everything take its course. Put on the music or let the dancing evolve in silence.

Anybody who feels ready gets up, dances their solo, finishes, sits down. The next one gets up, dances, sits down. Then the next one. And the next. When no one else gets up, ask if some prefer doing their dance two at the same time, or three. Or everybody at once, which also is a nice way to conclude the exercise.

Still another alternative is to let everybody sit spread over the floor. When someone gets up and starts dancing—or starts sitting or lying on the floor – everybody watches for a while and then joins in, dancing along to the end and sitting down together with the solo dancer. Then the next one gets started and everybody joins in after a while.

By the time everyone has done their solo, the atmosphere is dense and full of wonder. Sometimes there's exhilaration and laughter, sometimes deep affection and tears, also in teachers witnessing these small miracles of dancing and becoming. Allow space for discussion and comments. Ask for concrete observations and sensuous experiences of particular movements, images that arose. Ask how it felt to dance a solo invented all by oneself. Or how it felt to join in and dance along in someone else's work. Ask if anyone had butterflies in their stomach, and bring up the question of risk-taking and its outcome.

Sharing Solos and Creating a Joint Choreography

In the next step the soloists instruct their dances to each other, a stage in putting it all together into a joint choreography. Let them work together in pairs, trios, quartets or in a mixture of constellations. Each dancer shows their solo carefully bit by bit and is equally observant about the others' work. Then the pairs, trios and quartets put the solos together. Along the way there is plenty of space to further expand and interpret the material of movement.

Go round and highlight some particular quality or way of performing a step or movement. Ask for suggestions for further variations. Make use of where in the room the different small groups are and propose a spot from which to start their assembled dance piece. Two dancers standing by a wall can start right there and aim at finishing on the other side of the room, and on the way there alternatingly stick close to each other and drift apart, playing with distance. A trio may start facing the window, backs turned to the room, standing in a line side by side, then slowly dance backwards, turning around while doing one of the movements, either all three at the same time or one by one, then spread out rapidly in the room doing another movement, go on dancing in their own spot for a while before reuniting side by side, and so on. A third group may start in a circle in the middle of the room, move away and back towards each other while making the circle spin round, move longer and longer away from each other out to the corners of the room, and so forth. One pair may start in one corner and dance along a diagonal towards the opposite corner, in a roaming movement back and forth, alternatingly moving close and apart from each other. Or they start off in opposite corners and move all over the place until they finally meet in the centre.

Decide on an order for the assembled choreography. As one constellation finishes their part the next one will start. Ask the last group to show their final sequence of movements; the others look closely so they are prepared to join in in a collective finale.

Ask for silence and concentration.

> At your signal everyone dances their original solos simultaneously. Then they go to their small constellations at their respective starting spots. Put on the music. The first group starts their dance, finishes, sits down or backs out. The next group does their dance, then the next one, and so on. As the last group dances and gets to their final movements all join in and dance along to the end. All can finish following the last group, or do a final improvisation, or dance their original solos one last time.

This setup is one example of structuring a joint choreographic work. It can be varied to taste. One way of reinforcing the sensation of creating a space for one's own dance is by everyone placing a pair of shoes, a garment, a toy, a piece of cloth or some other personal prop out on the floor. In silence or when the music starts everyone enters the floor and goes to their place, ready to start the dance.

Have a meta-talk and reflect on the young dancers' and choreographers' delicate work. Draw attention to various aspects of movement that filled the floor and the room, and images and impressions that emanated from the dance. Help put it into words and assist in describing how this magnificent joint choreography came about, starting from everyone's solo dances that were further boosted and refined through a tight and inventive cooperative work.

Dance and Education: Aesthetics, Ethics and the Existential

Meta-talks are not about assessment. They offer an opportunity to refer back to the dance and choreography work and further enhance kinaesthetic awareness and literacy by helping the young creators to describe all there was to *see* and *sense* in proper words. Time after time Greene points out that encounters with art forms, in school as in life at large, help us to become more present and aware about what there is around us, help us to sharpen our gaze, 'to see what there is to see' in an art work, a dance piece, a poem, in everyday life and in life as a whole.

In terms of aesthetics the talks prolong the intense presence the dancers created and the heightened perception with multiple senses. The more we take in, the more we discover there is to take in. In kinaesthetic terms specifically, perception widens from matter-of-fact *observing* bodies in movement to *experiencing* others' movements with our own bodies. Not only do we *notice* what factually unfolds in front of us on the floor. We sense the qualities of movement carefully worked into the choreography: shifting spatial distances, speed contrasts, a minimal movement in a body part, a pending swing of a torso before a muffled crash on the floor.

As spectators, we also 'see' all kinds of pictures emerging: of weightlessness and spiritual beings, of sojourns in deep water, in a fog, at dusk, in fairyland; we remember a friend, a relative, a gesture, a life occurrence small or big. All this the young dancers create in front of our eyes, as 'gifts' we receive through our senses. On an empty floor being filled with movement. What we see are agentive subjects, adding something to the world that was not there the minute before. What they bring in is received in a welcoming community of others who deal with their initiatives with care and enrich them. Dancing together they evoke unforeseeable worlds where everything can be different. Still, this is the real world, the social world where we encounter each other, where we are with each other, all capable of acting upon our freedom of movement and our freedom at large. What we witness are new beginnings, acts of becoming.

Assessment and grading jeopardises all this. Judgment risks reducing the indispensable value of young peoples' actions and initiatives for their education, opening up spaces of becoming. Their creative work must be nurtured in a spirit of a welcoming community. It must not be laid in the hands of others to make judgments about. That would be to restore the subject to the realm of objectification, a relapse into a reducing, de-subjectifying and de-aestheticising view on school and education. It would deprive the aesthetic of its ethical and existential relevance, ultimately its significance for a democratic education.

During the compositional work from the solos to a joint choreography numerous small miracles occur, of dance and creation, of art and becoming, of aesthetic experiencing and ethical receptivity. These are superb moments. They touch the core of what dance art is about and what dance in education can remind us of what education is all about: trying to enter into dialogue with the world and to be in the world with others. The hesitation and resistance that may arise, weighing whether to take the risk to emerge with one's dance or not, ultimately touches upon what dance, art and education is all about. To break into the world as a subject or to abstain.

Security and trust are created not for the sake of avoiding moments of resistance or challenges the child or student may encounter in the work. They serve as an incitement, a support to take courage and go into what Biesta calls 'the difficult middle ground': between backing out and refraining from the effort and dashing ahead and tearing down.

A child storming in like a ball on a bowling alley may harm themselves or someone else, or eventually destroy the dance. The child needs support to see grains of gold in their movement pattern and wash them out through a larger sense of control of their physical and sensuous assets. This goes for work with clay or woodcraft, in singing, guitar playing or drama play. In this respect working with dance shows limits for self-expression and one's own desires, in favour of the intersubjective interplay and what is desirable for the common good.

Similarly, strategies are needed to invite the child or pupil who is anxious and wrestles with their confidence or shyness to add their work to that of the collective. Working with dance shows them a pathway to be on speaking terms with the world, to step forward with their unique senses, their imagination, their aesthetic resources and their own unique grains of gold, indispensable to the world that is everyone's.

Part III Letting Dance Teach Language, Mathematics and Natural Sciences

In the following chapters the dance teaching method is methodically integrated in the teaching of English, math and science. In the three-stage teaching process presented the point of departure is dance, followed by meta-talks, and a concluding stage formalising the subject content dealt with.

Dance is the starting point, or rather the avenue through which the other subject matter is 'received'. Dancing, the child or student opens up to subject matter through the senses and is offered opportunities to (kin)aesthetically experience the subject matter at hand. In meta-talks the subject content discerned in the dance activities is discussed and named in appropriate subject vocabulary, providing opportunity to become cognitively conscious of the content. Finally the participants help out to formulate idiomatic and grammar rules, definitions and equations, principles and axioms.

The dance teaching method is presented as *aesthetic teaching*, foregrounding the work of the senses in the process of learning. As elaborated on in the theory section with reference to Roth, cognition and knowledge acquisition are preceded, and made possible, by the sensuous. Many of the dance exercises are borrowed from the dance teaching method.

As dance meets the other subject, the child and student are offered an abundance of opportunities to enhance their kinaesthetic literacy—to 'think' like a dancer—while simultaneously develop their general language, math or science literacy—to think like a linguist, a mathematician or a scientist. The avenue is the (kin)aesthetic experience of language, mathematics and natural sciences.

Introduction

Dance, language, math and science are four subjects, fields of knowledge and proficiencies, literacies, and ways of perceiving the world. Whereas dance as an art mode deals with sensuous exploration of movement and artistic creation much for its own sake, beyond the predefined, the rational and the cognitive, the subject content in the other disciplines is so to speak determined. It is 'out there in the world', social and natural, waiting to be discovered, explored, interpreted and comprehended.

In the method setup described below, bluntly named *Letting dance teach*, dance shows the way to the worlds of language, mathematics or natural sciences. In the course of the dance exercises, dance brings out various subject content and becomes a window opening onto that particular field of knowledge as part of the social and natural world.

Phrased in Biesta's words, dance emerges as the archetypical *form* of teaching: *to point* at something in the world that may be good, worthwhile and meaningful to pay attention to. Dance also appears as the archetypical *object* of teaching, as *picture*, visualising content, making aspects of language, math, physics or chemistry 'receivable' by the senses. The archetypical *modality* of teaching, *to explain*, evolves as a collective effort by the participants, guided by the teacher, discussing, naming and defining the subject matter revealed in the dancing in subject appropriate language.

Teaching dance integrated in the teaching of language, math or science also points at all the language, math and science aspects of dancing, and at all the aesthetic inherent to language, math and science. Into the bargain the pupil's general literacy and language development are stimulated. Handling subject specific terms and concepts related to dance, math, science and linguistics further opens up for different ways of naming and looking at the world. By that means the method encourages critical thinking and a broader liberal arts education—to see connections and to see the world in relief.

Dancing the Didactic Triangle

As the dancing starts, a *space of learning* (Marton & Tsui, 2004) is established, consisting of intense interaction between the teacher, the pupil and the subject content—the *didactic triangle*. In the method it is firstly a space of aesthetic experiencing, a space where worlds of knowledge reveal themselves aesthetically, to the senses. The receptive work of the senses precedes cognition, making cognition possible, and lowering thresholds to subject fields many young people perceive as difficult.

Chapter 4
Dancing Through the Alphabet

P. Moerman, *A K-12 Dance Method to Teach Creative Dance, English, Math and Science*, https://doi.org/10.1007/978-3-031-93156-7_4

Kinaesthetic and Linguistic Literacy in Interplay

Dancing through the Alphabet is a dance and English teaching method aimed at children's and students' kinaesthetic literacy development, interplaying with their English speech and written language development. The starting point being dance, several dances described in Chapters 2 & 3 on teaching creative dance reoccur and are referred to.

The underlying idea is that these dances, already as they are without any arrangement, can 'teach' us language, pointing at the language in the instructions and how they are danced. The teaching-as-pointing modality is reinforced by applying minor emphasises and angles, drawing the dancers' attention to specific language content. In the meta-talks following the dances speech and writing aspects are highlighted, verbalised and formalised in idiomatic and grammar rules.

Dancing through the alphabet is a joyful and interactive teaching and learning method I have practiced with thousands and thousands of children in preschool, elementary school and after-school activities, in inclusive education and in introductory classes for newly immigrated children and adolescents.

Helping Tools

Helping tools and props are used such as coated letters laid out over the floor, flags with names of dance steps, fabrics, fans, tambourines, castanets and other instruments, hand puppets and enlarged writing tools. See these as inspiring examples and add other artefacts at free choice.

Setup

As the children enter the room the characters on the floor give rise to spontaneous comments, stimulating spoken language. The musical instruments, flags and colourful objects announce a playful activity on the schedule, filled with colour and motion. The dance activities, focusing on movement and kinaesthetic awareness, create a common platform for the participants to engage with dancing and language learning. Communication thresholds due to differences in word-stocks and majority language skills are lowered. Everyone gets a chance to indulge in movement, music and the joy of dancing. Those who are hesitant and prefer to wait can just sit by the side and watch and listen. Invite them now and then to join in on the dances or the discussions.

The attention is turned to what letters look like, how they sound, in which names, words and verbs they occur and are put together, in synonyms and antonyms, in prosody, in rhymes and alliterations, all of which are embedded in body and movement.

Each part of the session starts with dance, followed by a meta-talk in which various linguistic aspects are highlighted and different areas of literacy development are dealt with.

What dance does in all this initially, besides creating opportunities to advance in dance skills, is making language visible and audible—and danceable. The dances *point* at language, *teach* language. The verb *turn!* is heard in the instruction (the verb's imperative tense) and is simultaneously carried out by the dance teacher and the children *turning* (the verb's gerund form). The meta-talks give an opportunity to find the words to describe what a *turn* is (the noun in the indefinite mode), how we did *the turn* (the noun in the definite mode) or how we *turned* (preterite tense) while someone might mention how our arms *were stretched* out (preterite passive voice) or that our arms were *stretched* (perfect participle). Which specific language aspect or linguistic term is touched upon and named depends on the context and the group.

The meta-talks can be structured by applying mapping methods. Below, key literacy development areas (Eriksen Hagtvet 2006) are used: general linguistic proficiency; situation-dependent language; linguistic awareness; and written language development including experiences of writing and reading. Examples are given following some of the dances described.

The lcsson setup and sections presented below can be variously modified and expanded.

Names Gathering

Everyone goes to 'their' letter: the first letter in their name. Go round and acquaint yourself with everyone. Emphasise the first letter in every name. Mention if here is a Caspar or a Clara on a C; the other day we had a Kaspar and a Klara on a K. And here is a group of four on the A: Aisha, Anisa, Anna and Adam!

Ask the children to point at your letter, turn on some background music, take a tambourine, and go and stand by your letter. Ask them to do something that starts with your letter: *Point at Paul!* Or *Wave at Wilma!* Shake the tambourine, lower your voice—are you ready …?

A Literal Warm-Up

Music: atmosphere creating, without driving pulse, e.g. Miles Davis: Concierto de Aranjuez by Joaquín Rodrigo.

Pick three letters, for example S, G and H, or K, V and B. Guide through a warm-up with verbs on each of the letters.

Rhythmise your voice, with help from the tambourine to reinforce beats and rhythms, stops and ritardandi, while giving instructions and dancing along. Create a continuous flow. Vary the movements in the room: crisscross or in a zigzag, in straight or curved lines, in a circle, on the spot, with small or large displacements. Vary time—fast, stalling—and force—jerky, smooth, energic. Vary use of the body—from big movements to smaller, engaging the whole body or only the hips, arms and legs, mouth or eyes:

> Stand still! Slowly, slowly stretch your arms up in the air … Spread your fingers! Let your arms sink down. Stand on the tips of your toes. Press the sole of your feet to the floor. Stomp and stamp! Shake it! Shake your bodies. Smile! Stomp and stamp, shake and smile! Shuffle and shamble! Stand still! Steal away on tiptoe … Sneak behind Sandra … Slide and slither behind Serena … Slip and stumble! Sit down! Stand on your knees. Sit on your heels! Stand up! Shake your shoulders! Stand still …Shake your shoulders, shiver and shake! Stop, stay where you are! Stand stiff like a stick! Slack off! Stand stiff like a popsicle stick! Slack off, soften and melt in the sun, sink down on the floor!
>
> Stand up! Skip! Skip round and round in a circle! Stall! Spin around! Spin in the other direction! Sneak behind Siri! Sneak circling around Siri! Sneak in front of Siri! Stop! Skip, everyone, skip behind Siri! Slip and fall!
>
> *Put in a triple tizme:* Swing it and swing it and swing it and spin! *Sway your arms from side to side three times in a pendular movement, spin around on the fourth triple time*: Swing it and swing it and swing it and spin! …
>
> Squat down. Sit on your heels. Sit down on the sand and get a tan in the summer sun!

Put the tambourine aside, shift to castanets, spoons, claves or any other percussion instrument, hold on to a steady rhythm, change letter:

> Go to the G, go to Gustavio's G! Gather around the G! Gather and gaze at each other! Gaze! Gaze at Gala! Gaze and giggle! Go around Gilbert. Gaze at Gilbert! Gaze and glare! Go around and greet each other! Let's go! Go faster! Gallop! Gallop like Gandalf's horse! *Create a rhythm with your voice.* Gallop, gallop, gallop, gallop!
>
> Gather in the middle of the room. Gaze at each other! Groan! Growl like a grizzly bear! Grunt like a pig! Gurgle! Get the giggles! Get going! Go find Gustavio. Gang together with Gala and Gabriel! Go around over the floor, helter skelter, helter skelter. Helter … Hey! … Hold on … Hold your horses …

Change instrument, take a hand drum, shift to a new letter. Stroke your fingers or nails over the drumhead, make a hissing sound. Lower your voice.

> *Puff*: Helter skelter, helter skelter … Ho ho ho … Halt! Hold on! Where is the H, Hanna's H? Hunt out the H! Hanna, help us find your H! Hang on to Hanna! Hurry to the H! Here is the H! Hop, hop, hop! Hop on one leg! Hop on the other leg! On the first leg! Hop with both feet together! Hop and hop around, helter skelter, helter skelter! Halt! Clear your throat, hawk! Hang in there! Hasten towards the door, hurry, head over heels! Halt! Hold your horses! Hug your neighbour. Hum a happy tune together. Say hi to Harumi! Hush! Slip, hobble and fall.

Turn off the music.

Meta-talk on Dance and Language

Take a breather sitting among the letters on the floor. Allow time for spontaneous comments on the dance, let the dancers share their reflections. Be ready to write words and lists on the board or flip chart, or ask someone to pick up a marker pen or a big toy pencil and write. Direct the attention to both the grammar and the idiomatic richness of words, synonyms, sounds and rhythms—*as easy as pie*.

Help find the words to describe steps, movements and the sensation of dancing as well as the lively interplay engaging the group. Help retell and put words to everything we did from the warm-up. Ask if anyone remembers which letter or whose letter we started with. *S as in Selma*. What did we do on the S? *Stamp!* Yes, I said *stamp*, and everyone *stamped*. *Shake!* Right, I said *shake*, and everyone *shook*, and *smiled*. Which body parts did you shake? Shoulders. Heads. The whole body. *Shudder!* Shudder and shiver, everyone!

We sneaked on our toes! Right, I said *sneak* and everyone *sneaked* on tiptoe in a long *line* behind Sandra. We *skipped*! We sure did the *skip step*, a basic dance step. I said *skip*, and everyone *skipped*. How does the beat on the drum sound when you do the skip step and the feet touch the floor? *Skip! ... a-skip! ... a-skip! Stand still!* I said *stand still* and everyone *stood still*, stuck and fixed, immovable, straight backs, stiff as a stick, stiff as a poker.

Ask if anyone knows anything you can do on the S that we did not do. *Swim!* Show us how you swim! Swim along everybody! If I had said *swim* then everybody would have *swum*. *Sleep!* If I had said *sleep* everybody would have *fallen asleep*, and maybe *slept* the rest of the hour. *Swoon!* Does everyone know what *swooning* is? Ask the one who said *swoon* to show and ask everyone to imitate. How do your bodies go when you faint and *fall into a swoon*?

Someone may say *cycle*! Take every opportunity to enhance linguistic awareness of phonemes and characters. *Cindy cycles to the circus* does sound similar to *Sara stomps and stamps*, but is written with C's and not with S's. C is tricky when you consider listening to: *Carole was cool who could climb up that cliff ...*

In these meta-talks the sensation of moving as instructed by the verb's imperative mood is linked to an awareness of the phonemes in the verb and how letters build words. Discussing language offers chances to crack the alphabetic code, link sound to written symbols, analyse the content and the ring of words and sentences. Connecting the letters in one's name enhances self-anchoring and engagement. Reinforce the ability to tie to written language and the visual by writing lists of verbs on the S, C, H, G on the whiteboard—facilitate experiences of 'playing writing' and 'playing reading', social activity for which the dance cleared the stage and in which all learn from each other.

What did we do on the H? *Hop!* I said *hop*, and everyone *hopped*. How did we hop? We hopped around on one leg, hopped on the other leg, hopped on both legs. On which letter does *gallop* start? *G!* Like *Gala, Gustavio, Gillis* and *Gabriel*. How does the rhythm of the *gallop step* go? *Gallop! ... gall-op! ... gall-op! ...* What more did we do on the G? *Giggle!* I said *giggle* and everyone *got the giggles*. *Go around*

Gilbert! Yes, I said *Go* and everyone *went* circling around Gilbert. *Go* is a very common word. You hear it in all kinds of phrases: *Let's go. Go on. What's going on? We're on the go. Get going. Go ahead. Go back. Go away. How did it go?*

Let's go for a walk, a big walk.

Walk!

Music: rhythmic, stirring, e.g. Bizet, Carmen: overture; see detailed description in 'The Whole Body Walking' in Chap. 2.

Let's walk with the whole body: heads, hair, elbows, arms, shoulders, stomachs, hips, legs, knees, and of course feet and toes. Walk crisscross all over the floor, change directions in sharp angles when you meet someone.

Show a drawing with straight, cornered lines, and one with smooth winding turns. Put on the music. Beat the rhythm on a hand drum. Vary and adapt the choreography to the music:

Walk and move your whole bodies, let the arms walk along, hands, shoulders, elbows, torsos, bellies, hips, knees, legs and feet, the whole sole of the feet touching the floor.

At the change of melody: Walk on the tips of your toes, tip-toe, tip-toe, walk on tiptoe, tiptoe, tiptoe. Walk on your heels, heels, heels. Walk on your toes, on your heels, on your toes, on your heels. Toes, heels, toes, heels.

Change of melody: March like a soldier, one, two, one, two, one, two ... *New change of melody*: Walk on the inside of your feet, on the inside, can you find the inside of your feet? Walk on the outside of your feet, the outside, the outside. On the inside, the outside, inside, outside, in, out, in, out ...

New change of melody: Move like spaghetti, fling out your arms and legs, feet and knees, your knees give way, your head goes hither and thither, hair as well, hips are slack, legs like jelly, move like boiling spaghetti. Walk with that whole loose body.

In the stop in the music: Freeze!

Change of melody: Sneak on tiptoe ...! *On the timpani beats in the music*: Stretch arms up! Down! One arm out to the side, the other arm out to the side! Quiet feet, quick feet, tiptoe to the walls. Turn your backs to the wall, stretch arms to the sides, tiptoe sideways along the wall, all in the same direction.

Stop! Sneak forward on the tips of your toes, to the middle. *On the timpani beats*: Arms up! Down! To the one side, to the other side! Quiet feet, quick feet, on tiptoe to the walls, turn your noses to the wall, arms out to the sides, on tiptoe sideways along the wall.

Stop! Sneak backwards on tiptoe, towards the centre, stretch arms up, down, to the sides. Keep your arms stretched out to the sides. *The music staggers*: Turn, spin around, faster, faster, still faster ... *The music stops abruptly*. Sit down.

Meta-talk on Dance and Language

Talk about the dance. Did you get giddy? We spun round faster and faster. Talk about how our whole bodies got going in the dance. Name body parts. Walking on the tips of our toes, on our heels, on the outside and the inside of our feet. Talk about how we moved round the room, and where, about the feeling when we sidled and slid along the walls on tiptoe. How big and empty the floor got when we all had our backs to the walls, before we slowly crept forward on tiptoe taking the floor again.

Talk about the force in the movements, the difference between the frenetic tiptoeing and the stealthy sneaking steps. About the sensation of the whole body creeping. About the labour of the shoulders, the swing of the hips. Talk about the time aspect, the turns going faster and faster, the spaghetti cooking and boiling. Connect to the rhythmical drive in the music.

Can you do the fast and forceful movements to the calmer music, and the other way around? Help them to put words to everything, to 'think' dance and expand *kinaesthetic awareness.*

Go into idiomatic language. *Walk the line. Walk the walk. Walk on eggshells.* Marching like a soldier or wobbling around like boiling spaghetti enhances *linguistic awareness.* A word can be a human or a thing. Or something you can do, or dance. In dance we can do everyday things in an imaginative way.

The Birds' Dance

Music: free flow with rhythmical parts, e.g. Radio Tarifa: Canción Sefardí; see detailed description of the steps in 'Arms Like Wings to Fly With' in Chap. 2.

Everyone squats down, spread over the floor.

> Imagine we're birds, our arms are wings. The birds sit in a tree, sleeping. It's cold and dark, morning hasn't broken yet. The birds wait for dawn and light, for the sun to rise and warm the air. Our sun comes with the music.

Put on the music. Squat down, stretch out your arms gently, fold them behind your back.

> Here comes the sun. The birds stretch their necks and rise a little, feeling the air with their wings … is it warm yet …? No, brrr … Crouch down again. Let's check again, feel the air with your wings … Oh no … Still cold, chilly morning. Huddle up again. Third time lucky? Yes! … Now the air is nice and warm. Rise and shine, let's stretch our legs and spread our wings. Big strong wings, ready for a great tour up in the skies. Feet on tiptoe, rapid and silent feet, this is a silent dance.
>
> Off now and fly, lift your wings, spread them out, soar through the skies.
>
> Stall. Spin round, in one direction, in the other direction, keep your wings spread out, feel the air under your wings. Stop … Do big wing strokes on the spot, high up over your heads, stretch your bodies, up on the tips of your toes, straight legs. Lower your wings, squat

down, wingtips touching the ground, up again, straight legs, down again, up, down. Biiig wings, strooong wings.

Fly away, sail up into the skies on the air spirals. Fly and flap, flutter and flop, forward, backwards, to the side, crisscross among the other birds. Stop, point one wing down to the earth, the other wing up in the sky. Spin round, look down, see house roofs, people down there, cows, horses. Switch, point the other wing up in the air, look up to the clouds, spin round, do you get a little dizzy?

Off again, fly on. Big wing strokes. Fly crisscross, air acrobats, forward, backwards, sideways, round, round, low over the ground, soar high up in the sky. Make small eager movements, like small passerines. Quiver your wings like hummingbirds, hover, hang poised in the air, fly backwards. Make mighty wing beats like a loon.

Gather all the birds at one side of the room, fly in front of the flock.

Fly forward. Slow down on tiptoe. Forward again, break again, fly, stall. Fly away everyone, crisscross. Fly home to the tree, the sun is setting. Land on the branches, squat down.

Hop up, switch branch, land, change branch again, change again, again, again. One last time, change branch, land sitting on your heels. Fold your wings, shake your feathers. Huddle up for the night, go to sleep. What a great day for flying it was.

Meta-talk on Dance and Language

Talk about birds' way of moving, the feeling of dancing feather-light and soundless, the sensation of cleaving the air, feeling a fan against the cheeks during the flight, feeling the air under the wings in a turn. Talk about balance and the sense of holding back a movement. Ask for comments on the work of the shoulders moving arms like wings. On the labour of the thigh muscles stretching and bending when we crouched down on our heels and rose again and did our huge wing beats. Our thighs got to work still harder in the end when we switched and jumped from branch to branch.

Dancing makes your body and muscles strong. A dancer needs to work out as hard as a swimmer or a football player to get strong and supple, and to build stamina. Discuss other muscles and joints in the motor system that can bend, stretch or twist. Demonstrate a twist with a hand puppet or a puppet on a stick. Talk about symmetrical movements, or asymmetrical, like when we span round one wing pointing to the earth and the other up in the sky.

A dance like this one along with the instructions unfolds in itself as a *here-and-now-dialogue* (Eriksen Hagtvet 2006). New *terms* are made comprehensible by saying and dancing them in the shape of a *story with a continuity*. The birds sit on their branches in the tree in the chilly night, they are asleep, they wake up, the day is dawning, it turns lighter and warmer, the sun comes with the music. The words paint an image of what the young dancers live through in fantasy. They are encouraged to check how it *feels* in their 'wings', how *opposites* feel between cold and warm, high and low, forward and backwards, up in the sky and down to the ground, bending and stretching, in one direction, in the other direction.

Naming and *emphasising* the words and *dancing* at the same time –flying and flapping, backwards, forward, to the sides—creates conditions for *concept stimulation. Morphological awareness* is awakened by word forms and word formations like *arms like wings*, *the wings*, *wing strokes*, *the one wing, the other wing*. All of these are experienced with the senses and in imagination. The *recognisable theme* helps the children explore their curiosity. The birds are shy at first but emboldened as they sense the weather, they start moving slowly and then go the whole hog. Invite and encourage the children to tell and *word* their *inner stories* and *associations*.

Discuss linguistic aspects. Make lists of words on the F, birds flying, flapping and fluttering with their feathers making the air fan. Ask for more words and names on the F and add to the list.

Bodies Shaping Letters

Which letters do you know? Let's do them with our bodies. Let's start with the hardest of them all …

> Stand like an I!

Go round and draw attention to how each one makes their I in their own way:

> Here's one with a dot on it. Here's a long drawn-out one straight like an arrow. This one is the smallest I I've ever seen, I almost tripped over it. And here are two small I's hanging together and hugging. This tall I is choking with laughter.
>
> Can you make your I bigger? Everyone rises to their toes. Even bigger? Arms stretch up in the air. Can you make your I tiny little? Everyone huddles up like small dots.
>
> Stand like a T. Stand like an X. Like an I … like a T … like an X! Like an I … like a T … like an I … like an X … like an X!
>
> Lie down like an L. Sit up like an L. Sit like a V. Lie like a C.

Go round and point out all the inventive solutions. One child once proudly showed how many X's they could shape with the body: standing legs crossed, feet making their own X, arms crossed and five pairs of fingers too. I counted eight X's. The child smiled and said I missed one, looking at me knowingly … squinting … *I'm cross-eyed!*

> Now find a friend and work together two by two. Do a little *i* lying.

Direct the attention to all the different solutions again.

> Make an H! Point out all the solutions. Some are lying down, others stand, some are on their knees, some hold each other, leaning backwards, bodies bending over. Say another letter for them to do two by two. P!
>
> Now work three by three. Make an E, an R, a K. Work four by four. Do an M.

Go round each time and point at the imaginative solutions.

Stop by a group of four. Isn't that a W? … Didn't I ask for an M? …

Now everybody together: do one single big lying I, the longest I you can make in the room.

If there is too much chatting going on, count to ten, or five. The bodies eagerly fall into the lengthy I. Sometimes everyone lies down in one single or a double line, or a triple or even a quadruple. Sometimes they lie as links in a chain, the one's head on the next one's belly. One child or a bunch of them make a dot by the one end. It happens a dot is put on the I at both ends.

Now finish by standing up and making one big O all together.

Everyone is standing in a circle holding hands.

How can you make the O bigger? Everyone takes a step or two back. Ask two of them to step out of the O and turn it into a Q. Do you know anyone who's name starts on a Q? Queenie! Quentin!

Ask the two to join the O, join in yourself, let go the hands, sit down.

Meta-talk on Bodies Shaping Letters

Ask for comments on all the letters they shaped in their own ways, alone and with others. Let them *retell* and *relate* and so practice their *linguistic proficiencies*. Connect to the plastic coated letters on the floor. *Concept stimulation* is intensified as the children hear and experience the letters sensuously with their bodies, follow instructions and work in small groups. By shaping the letters with their bodies and by naming them, they get further opportunities to discover the relation between *phonemes* and *graphemes*, without using pen and paper. Talk about the *appearance* of the letters: are they straight, curved, round, long, edgy? This theme is expanded in the next dance.

Picking Fruit in the Alphabet

Music: long piece, calm, melodical, e.g. Gipsy Kings: Passion, or Ravel: Boléro; see detailed description of the movements in 'Spanish-Inspired Dance: Picking Fruit under the Orange Tree' in Chap. 2.

Everyone is sitting in the big O-circle.

Sit in tailor position, legs crossed, like an X. Do you know anyone with an X in their name? Xavier! Xerxes! Alexandra! Why do we call this sitting position the *tailor* position? What does a tailor do? Sews clothes. Why, do they sit and sew like this? Where did they use to sit like this? On a table?! Discuss how tailors handle large pieces of fabric.

Now let's pick fruit. One fruit for every letter in the alphabet. What is the alphabet? *ABC. All the letters*. Any old how? No, in an order. In *alphabetic* order. In English from A, B, C to Z. In Greek from *alfa* (α), *beta* (β), *gamma* (γ) to *omega* (Ω), in Arabic from *alif* (ا), *ba* (ب), *ta* (ت) to *ya* (ي)—'*ya bata*', a little duck with two feet …

Put on the music.

Make a basket with your arms, like an O, gracious arms, bend your upper body slightly forward and back. Continue the arm movements in a wide arch in the air and behind your back. Feel if the small of your back is warm from all that dance.

Let's put all the fruits we pick in the basket one after the other.

What is the first letter in the alphabet? A! Do the basket, feel behind your back. Lift one arm, stretch it up in the air, pick an … apple! Pick the apple high up in the air, a smooth and gentle movement from the wrist, put it into the basket. Move your hand to your hip, lift the other arm, which letter comes next … B! Stretch up your arm, pick a … banana! Rip down the banana, put it into the basket, let your hand rest on your hip, the other hand rises. Then comes … C! Pick a … cherry! put the cherry into the basket. After the C comes … D … we pick …? Dates! Pick some dates from different spots with both hands. Do the basket. Which letter is next? E! … Elderberries!

Vary the size of the movements and their quality in arms and hands.

Fingers pick small blueberries, cherries, dates and elderberries from trees or bushes. Hands splay to pick a soft and yummy mango. Big arms slowly bring down a heavy watermelon.

Help out with fruit names when needed. Cheerfully welcome suggestions like eggplant, elephant or Ella on the E or a zebra on the Z—down into the basket among the fruits those sweet black and white stripes. Prioritise the flow of dance, add more fruits on all the letters at a later opportunity.

Now the fruit basket is filled with apples and figs and oranges and perhaps a xylophone and a zebra or Zorro. Do the basket again, a huge basket like an O, hands behind the backs, back again over the floor, sweeping under the basket, taking a grip under the basket. Lift the basket and put it on your head. Let go with your hands, hold your palms together in front of your body, lift your eyes towards the basket, balance it on your head. Try to come up on your feet, keep your balance. Lift the basket from your head, lower it to the floor, toss it up in the air so all the fruits fly out and fall down on the floor.

Now let's mash the fruit and make jam.

Stamping and Jamming

Music: rhythmic common time, upbeat, disco, e.g. Gipsy Kings: Lama bada; see detailed description of movements and rhythms in 'Stamp the Beat with your Feet: Make Jam' in Chap. 2.

Change the music. Stamp. Hands on hips. Stamp an equal rhythm, one foot, the other foot. Start in a slow and steady common time, then half the time to fourths, doubling the stamps, on to eighths, keep stressing the beat on the one. Try triplets. Speed up the rhythm until there is fast rattling stamping:

Mash, mash, mash, stamp, stamp, stamp, forward to the middle of the circle, lift your arms, shake your hands and fingers, back out again, lower your arms, shake your hands and fingers, to the centre again and back, forward again, mash, mash, mash, and back, back, back.

Meta-talk on Dance, Fruits and the Alphabet

Review dance aspects. Help name the movements, stamps and rhythms, the picking, cupping hands, body parts like wrists and fingers; movement qualities such as smooth or jerky, gracious Spanish style, oriental, spatial aspects.

Talk about the alphabet. Create a *there-and-then-dialogue* related to the dance as a *here-and-now-dialogue*. Dialogue draws on listening and seeing interlocutors who create presence and entice an interest and a desire to participate, to dear to speak and ask questions. List more fruits on the letters of the alphabet, animals, flowers, common names and less common names. This dance is a playful way to acquaint oneself with *the alphabetic principle: the code* to all reading and writing.

Spanish Arm School

This is a progression exercise on picking fruit. For a detailed description see 'Picking Oranges, Bananas, Small Berries and Big Watermelons' in Chap. 2.

This reference to a specific style exercise from Chap. 3 on teaching creative dance is a reminder that any dance activity can show a way into language, as described in the method. Value the opportunity to introduce a specific dance or style you or any participant are familiar with. Instruct the group or let the familiar young dancer lead a session. Encourage the use of appropriate language, discuss and highlight linguistic aspects in a meta-talk following the dancing.

Body Parts and the Whole Body Dancing

Music: rhythmic, upbeat; see detailed description in 'Warming Up Body Part by Body Part' in Chap. 2.

> Fingers dance. Fingers and wrists dance. Fingers, wrists and hands dance. Elbows dance. Fingers, wrists, hands, elbows, shoulders dance. Necks and throats dance, heads, hair.
>
> Chests dance, stomachs, hips, bumps. Legs, knees dance, feet dance, over the whole floor. Fingers dance, wrists, hands, elbows, shoulders, necks, heads, hair, faces, chests, stomachs, hips, bumps, thighs, legs, knees, feet, the whole body dances. Dance freely, at your pleasure!
>
> Dance like Hava! Dance like Ali! Like Oscar, like Felicia!

Meta-talk on Dance, Body and Language

This run-through moving each body part in isolation is a common warm-up in dance or mime, adding up to the whole body moving. When transferring to everyone taking after one in the group dancing freely, the 'leader' gets a chance to test the limits of their own choices as their movements are mirrored by the others imitating.

The leader can do almost anything they like, tapping into their own flow and mood, and everyone just dances along.

Awareness of both dance and language skills are enhanced in the meta-talk. By using *situation specific* language and gesture the children get a chance to express *interior monologues* they carried on while dancing and to voice sensuous impressions from the kinaesthetic experience.

Galloping or Skipping Two by Two

Music: rhythmic 2 or 4-beat, e.g. Radio Tarifa: Rumba Argelina; see detailed description in 'Galloping or Skipping Two by Two' in Chap. 2.

Half the group stand in a row side by side at the one short end of the room; the other half stand in a row opposite.

Join one of the rows and show the steps with one of the dancers in the opposite row, galloping or skipping:

Hands behind the back, gallop/skip towards each other, meet in the middle, gallop/skip round each other a few turns, gallop/skip on to the other side.

Show once again so the dancer returns to their row. For progression:

The two meet in the middle, gallop/skip arm-in-arm a few rounds, the other arm up in the air in a right angle from the elbow. The turns gain momentum from the upstretched arms while they also help balancing.

Turn on the music. Everyone dances two by two, one couple at the time. The others clap the beat. When everyone has danced in pairs, hop in in one of the rows.

Dance towards each other, the rows meet in the middle, stretch arms up, wave and shout 'bye bye!', gallop/skip backwards, then forward again and back and forth a few times.

Finally everyone in both rows turns a quarter round so their shoulders are directed to those opposite—for example all face a window or a long wall.

Dance towards each other side-galloping, cross in the middle, continue all the way to the other side, back again, once more, a couple of times.

Horses Galloping and Jumping over the River

Music: duple or quadruple time, paso doble, circus music; helping tools: long piece of cloth, hand drum, castanets; see detailed description in 'Galloping and Jumping Over the River' in Chap. 2.

Everyone lines up behind each other at one of the short sides of the room. Hands on hips.

What do horses do when they are in a hurry? They gallop! Let's gallop like horses do.

Across the floor in the middle lies a long piece of fabric. Imagine it's a river. The water in the river is dirty, these horses don't care stopping to drink. They want to show how high and agile they can jump over the river.

First the horses gallop one by one towards the river, jump over, gallop on and form a new line at the opposite side of the room. As one horse jumps over the river, the next horse sets off in gallop, jumps, gallops on and joins the line by the other side.

Turn on the music, use the castanets or another instrument to reinforce and drive the pulse on. The horses take turns and gallop and jump until they all line up at the other side of the room.

Gallop back, jump over the river again, form the line as from the start. This time jump a little higher, you can do that. Gallop ahead and stretch out your arms as you jump. Jump as high as you can, feel how long you can stay in the air, fly over the river, like a winged horse.

Demonstrate galloping and jumping arms spread out like wings. Show a picture of Pegasus. Lie down full length on the fabric river and let the dancers hop over you, or put a basket or a cushion in the middle.

Galloping One by One, Two by Two, Four by Four, Eight by Eight, All Together

Progression of the previous dance:

The first horse gallops, jumps, gallops on to the opposite end of the room, turns to the left, gallops back along the side, jumps over the river and positions themselves by the line with those waiting for their turn to take off.

The second horse takes off galloping as the first one jumps over the river, continues galloping, turns to the right, gallops back along that side, jumps over the river and positions themselves by the line with those waiting, facing the first horse. The third horse moves like the first, the fourth like the second, and so on. Two opposite lines build up by the starting point, at right angles to the starting line.

When everybody has galloped, jumped and turned back one after the other, the first ones in each new line take each other's hand, the other hand on the hip, they gallop together, jump over the river, gallop on, split up and turn in their directions as before, gallop back, jump over the river again and position themselves in the back of the line they started. When the first pair jump over the river the next two get going, one from each side, and do like the first two. Then the next pair take off, until everyone has galloped and jumped two by two.

So it goes on. Now two from each line join in, take each other's hands, the outer ones keep one hand on their hips, they take off galloping, jump, keep galloping, split up two by two, gallop back in the same direction as before and go back to their lines. When one quartet jumps the next starts and so on.

When everyone has danced four by four, four from each side take each other's hands, the outer ones have one hand on their hip, and all dance eight by eight. Then sixteen by sixteen. Once we were more than thirty-two horses, once a full sixty-four.

> Finally all dance at the same time. They hold hands, the ones in the ends keep one hand on their hip. Now they get ready to gallop and jump like one single horse. Everyone puts one leg in front, paw the floor with their hooves, ready to take off. Go and stand by the fabric to check how they come dashing and jump over the river as one fiery steed …

Meta-talk on Dance and Language

Talk about the music and the rhythm of the gallop step. Talk about how everything went on and on non-stop, galloping and galloping in that steady cadence, like a heartbeat, one endless flow from the beginning to the end, unceasingly with new displacements and regroupings, from one side of the room to the other and back, one by one, two by two, more and more, until all danced as one horse. Discuss the distinct form of the dance, the importance of holding hands on the hips and galloping with the same leg ahead. Talk about the feeling in the jump, the sense of almost flying over the river with arms like wings—the magic of dance.

Talk about the importance of waiting for one's turn and of sticking to the gallop step despite the eagerness to show strength and speed and high jumps. Point out how smoothly they all danced the patterns and managed the whole flow of choreography, one uninterrupted cooperative effort which made all the parts fall into place and created clarity and elegance.

Discuss new terms that were introduced: jump in the middle, jump again by the sides, turn in one direction/left and in the other direction/right, and other *positional terms*. Leave space for own descriptions and explanations. This way *reading* and *imitating* are given another content; the young dancers can tell in their own words how they followed the instructions, and they can give own suggestions as how to gallop, jump and skip and move in still other ways like horses do.

Galloping and Skipping with Two Flags Flying

Music: duple or quadruple time, driving, e.g. Juan Peña Lebrijano & Orquesta Andalusí de Tanger: El Anillo; see detailed description in 'Dancing into the Room in One or Several Loops' in Chap. 2.

This dance is also suitable as a warm-up or restart after a break. The two loops can also come into the room dancing with the flags aloft.

> The dancers line up in two lines behind each other. The first one in the one file holds a flag with 'Gallop' on it. Those behind keep their hands on their backs. The leader with the flag gallops freely through the room, everyone behind gallops along. The first dancer in the other line holds a flag with 'Skip' on it and leads their line skipping through the room, the followers with their hands behind their backs.

Turn on the music, reinforce the dynamics beating the rhythm on a drum or tambourine. Stop the winding lines now and then at an appropriate moment in the music, and put in a movement on the spot, like an S, slithering like a snake.

Towards the end, the first line forms an inner circle side-galloping in one direction; the other line forms an outer circle side-galloping in the other direction, all faces turned to the centre. The dance in the circles continues. After a few rounds everyone turns facing outside for eight counts, side-galloping on without stopping, turns to the centre again for eight counts, outward four counts, inward four counts. Dance along in the outer ring, beat the rhythm on the drum.

Stop the circles, clap the rhythm together, then clap and stamp the rhythm, try different accents and phrases clapping and stamping. Move inward stamping and clapping, gather in the middle, close to each other. Back out again, spread over the floor, spin around arms stretched out. Spin and go down bending the knees, arms stretched out, sit down legs crossed, arms billowing, one up, one down like an S, up and down, the movements ebbing away while the music tunes out.

Meta-talk on Dance and Language

Talk about directions in the room, the two lines proceeding each with their dance step. Discuss the rhythms of the galloping and the skipping steps, and how they can be performed to the same music. Recall the two circles in the end, going in opposite directions while both are side-galloping. Connect with the words 'Gallop' and 'Skip' on the flags: the letters, the syllables. How do the verb words feel when danced as steps in the body? Keep discussing how we write and form words, linking phonemes to graphemes.

Sprinters, Sneakers, Stressers and Slackers

Music: 4 different tunes, from rhythmic to sneaking to rattling to stalling, e.g. Radio Tarifa: Las Cuevas, Conductus, Baile de Almut, Canción Sefardí.

We return to movements and verbs on the S while at the same time exploring an essential aspect of dance, the contact between feet and floor: whole foot, on tiptoe, on the heels, on the inside and on the outside of the feet. As dance educators often stress: *The floor is your best friend.* All movement starts from the floor and evolves in the interplay with gravity; working against it or surrendering to it—a basic experience of the body in motion.

Visual support may be added to this dance with pictures: four laminated drawings of the sole of the foot, one where the whole sole of the foot is coloured, one the toes, one the heels and one with the inside and the outside of the foot.

Divide the group into four, with equal numbers of dancers in every corner of the room.

Introduce the first group: the *sprinting people*. Show their way of moving around in the world.

They run and sprint light and jolly, *the whole of the sole of the foot touching the floor*. Sprinting is the most fun thing they know, they sprint limber and light-footed, forward, backwards, to the sides and round, round, round. They run everywhere, as if the whole world were their parkour scene. Now and then they hop up in a jump for joy, before sprinting along, whole feet on the floor.

Turn on the music. The sprinting people dance out on the floor sprinting around everywhere. Encourage energetic movements in the whole body. Remind them about the jumps and about the specific floor contact.

Sprint back home sprinting folks! A round of applause!

Introduce the second gang: the *sneaking people*.

These people sneak around slowly and carefully, *on tiptoe*. They sneak and creep with the whole body, stretching out their arms gropingly to all sides, lifting their legs in long strides, slowly moving, forward, backwards, sideways, round, round, round all over the place. They try to make themselves invisible and airy, make their body light, smooth and supple, like a cat or a tiger. Sometimes they squat down, they don't want to be seen, or heard, they keep still a while, rise slowly and go on sneaking, stealthy, invisible and soundless, always on tiptoe.

Put on the music. The sneaking people sneak ahead and take to the floor on tiptoe, moving sideways, up and down, filling all space between the walls, the floor and the ceiling. Whisper and remind them to squat down now and then, crouching and staying all still for a while before moving on.

Sneak back home, sneaking people! Applause!

Introduce the third gang: the *stressing people*.

These people are always in a hurry, always *on their heels*, stressing forward, backwards, round, round, round, they have no time to put down their feet, they labour frenetically with their whole body, head and shoulders shaking, elbows flying up and down, in and out, elbowing their way everywhere, hither, thither, they don't remember where they were heading, only that they are in a hurry, so they stress even more. Sometimes they stop and watch their watches, then hop up knees high and stressing further even more, all the time on their heels.

Put on the music. The stressing people rush out over the floor, jolting and stressing round everywhere. Cheer on, remind them about their heels on the floor, to stress with their whole body, to not forget the head, to check their watches.

Stress home stressing people! Applause for the stressing folks!

Introduce the fourth gang: *the slacking people*.

These folks shuffle about *on the inside or the outside of their feet*, forward, backwards, to the sides, round, round, round. They are heavy and tough, the whole body is slack and loose-limbed, arms hanging, heads hanging, necks are slack. They struggle with balance, they slide and slither, hither and thither without steering, wobbly and flabby. Now and again their legs give way under them and they sink down like shapeless jelly blobs, running out on the floor like a slime spot. With great effort they rise and shuffle on, bodies weighing like lead, staggering about, on the inside or the outside of the feet.

Turn on the music. The slacking people start dragging themselves along, sliding and slithering as good as they can. Remind them of their floor contact, cheer on when they slump down, to get up and carry on, to keep moving, shuffling or not.

> Now these slackers want to go home, drag on you lazybones! Let's put our hands together, they too deserve applause, they did as well as they could!

When each group of people have come out from their corner, danced in their designated way and returned to well-deserved applause, everyone gets a chance to dance and sense the different contact the feet can have with the floor. All dance in turn like the sprinting people, the sneaking people, the stressing people and the slacking people.

Put on the music, play a bit of each respective tune, cheer on and remind the dancers of the different modes of foot contact with the floor. As they finally sink down one after another they keep lying still and close their eyes. Proceed to the next dance to conclude the session and have a summarising meta-talk afterwards, or discuss the different folks' dances first.

Meta-talk on Dance and Language

Discuss the kinaesthetic experience of moving around in different ways with different degrees of contact between feet and floor, how this affects how the body moves as a whole. Discuss force, how the muscles tense when *sprinting*, *sneaking*, *stressing* and *slacking*, how much you can tone down to stillness or gear up to frenetic movement depending on the kind of floor contact.

The slackers ultimately crash and lie all knocked out on the floor. So it goes when forces peter out and you drop to the ground like an apple from its tree thanks to gravity: everything that has a mass is drawn down to the earth. Discuss the use of space starting from the four corners—the different people's 'homes'—out over the floor and populating the room differently before returning home. Discuss the different rhythms in the movements and the music. In movement and speech the dancers get to 'taste' the verbs sprint, sneak, stress and slack, all starting on S. Dance and language aspects intertwine in the meta-talks.

Structure the discussion of the linguistic aspects, for example according to Eriksen Hagtvet's (2006) four key areas of literacy development. *General language proficiency* is stimulated by the different instructions for the respective groups. Stimulation of spoken and written language draws on a distinct *here-and-now-dialogue* which here evolves during the instructing and the dancing, where the words used are easy to understand and given meaning in the context. The dance dialogue in each 'folk group' is reassured and supported by many peer dancers, and obvious to those watching their community of movement. The whole of instructions, dancing and discussions cater for a perceptive dialogue securing trust and space for those who might not quite understand to take the risk and speak up. The movement words starting on the S can be listed on the board in the verbs' imperative tense and in the gerund form ending on -ing, here also used as adjective to 'people'.

Situation dependent language is stimulated by relating to the activities just done—the core of meta-talks—from different aspects. Ask how the instructions were perceived and followed in the imperatives *sprint! sneak! stress! slack!* Ask for descriptions of how the sprinting people ran, the sneaking people sneaked, the stressing people stressed and the slacking people slacked. Ask about the sensation in the feet. Ask about the sensation of space when entering the room and moving in different directions, sideways between the walls and vertically between floor and ceiling. Ask about different speed and times. About the whole set-up—first listen, then wait for the music, then dance and get applause in the end. And now talk.

Linguistic awareness is enhanced by the *meta-talk* in itself—discussing language, instructions, demonstrations, clarifying gestures, how the movements are done, which body parts are involved: sneaking on tiptoe and moving slowly and stealthy, the whole body sprinting fast and springily, heads hanging and feet dragging. Abstract words like *stress* and *slack* are made concrete and corporeal through kinaesthetic experiencing. Word formations of letters and syllables such as *s-t-r-e-s-s + ing* or compositions like *sprint + people* are danced and stored in the bodies first. Dance and the (kin)aesthetic precede cognition and knowing. Movement, music expression, and language are integrated in a sensemaking context. The variation and richness of experiences cater for greater linguistic and kinaesthetic awareness. Experiences of *playing-writing* and *playing-reading* are scaffolded with the visual support of pictures of the sole of the feet. Repeatedly dancing movement verbs on the S reconfirms pre-understanding. Practicing dancing verbs and sounding them out precedes writing and reading. In this manner dance points out and teaches language. Meta-talks reinforce the pointing at the linguistic.

Cradling Waves

Music: sea rush, e.g. Radio Tarifa: Soledad; see detailed description in 'Rocking and Rolling Waves' in Chap. 2.

Everyone lies on their backs, spread over the floor.

> Close your eyes. Imagine you're lying on a sandy summery beach. It's a sunny summer day. We've taken a dip, a light breeze sweeps from the water over the sand where we lie.

Go round and fan gently with a big paper fan.

> Imagine you are the water, tiny small waves, ripples curling on the water surface in wind puffs. The wind increases, the waves grow bigger, swaying and rocking from side to side. The wind intensifies, the waves roll over, rocking back and forth in the wind.

Read Strindberg (1902), from *A dream play*:

> We, we, waves,
> are rocking the winds
> to rest!

Green cradles, we waves.
Wet we are, and salty;
like flames of fire;
wet flames we are.

Dousing, burning,
washing, bathing,
breeding, bearing.

We, we, waves,
are rocking the winds
to rest!

Lie still on the floor and relax, close your eyes, feel your heart pulse, think of everything we danced and discussed today, everything dance taught us about language.

Chapter 5
Dancing Math

P. Moerman, *A K-12 Dance Method to Teach Creative Dance, English, Math and Science*, https://doi.org/10.1007/978-3-031-93156-7_5

Kinaesthetic and Mathematic Literacy in Interplay

Dancing Math is an aesthetic teaching and learning method aimed at enhancing children's and pupils' proficiencies in dance and math, in other words their (kin)aesthetic literacy and numeracy development in interplay. As a dance teaching method, it is conceived within the creative dance method described in Chap. 2. Several dance exercises are borrowed from the chapter and referred to here, while others are designed to lead onto specific math content, such as double and half, multiplication tables, fractions, polygons and so on.

The method follows a three-stage procedure: dancing-discussing-formalising. More than a point of departure, dance is the avenue, staging a kinaesthetic experience: to sensuously experience movement and math, including particular mathematical notions and axioms. In a subsequent *meta-talk* the dance activity is discussed in dance and math terms, clarifying and properly naming the math subject matter at hand. Finally, the math content is formalised in equations, formulas and figures as they appear in typical math textbooks.

The set-up is aimed at creating conditions for the children and pupils to devote themselves sensuously to dance, math and language. Communication barriers due to possible differences in language skills are temporarily put aside for the benefit of giving rein to the work of the senses in both dancing and dealing with math.

The *meta-talks* referring back to the dance activity done can be structured by identifying any of Bishop's (1988) categories of mathematical activities, or as to which extent Gelman and Gallistel's (2009) principles are handled in the dancing. An example of such an analysis is presented below in a dance exercise for the youngest children, dealing with the circle.

In teaching and learning terms, dance in the *Dancing Math* method establishes a *space of learning* (Marton & Tsui, 2004): a space of interaction between teacher, pupils and specific math content—*the didactic triangle*. The dance exercises create opportunities to take in math subject matter through the senses, to visualise the young dancers' previous knowledge and conceptualisations in math, and offers keys to mathematical thinking—kinaesthetic literacy meets mathematical literacy. *Pointing* at math, *offering keys* to break mathematical codes, dance *teaches* math. Into the bargain, discussing dance and math aspects of the dance exercises stimulates, besides dance and math subject-specific language, also general literacy development.

The teaching method also offers an alternative approach to numeracy for the youngest: for example, in dealing with numbers and quantities. Instead of counting on their fingers, the children dance, and get the chance to kinaesthetically experience quantity and part-whole structures in the first natural numbers—a precondition to developing arithmetic skills.

Kinaesthetic experiencing thus advocates the encounter of mathematical thinking as *perception*: to see, hear and feel math, to let corporeal experience precede and cater for cognitive understanding—to let dance teach.

Helping Tools

The helping tools used include plastic coated numbers one to ten laid out on the floor—Roman, Arabic and Indo-Arabic digits each in their own colour—plastic coated numbers 1, 10, 100 and 1000, the irrational number ∞, eleven paper plates with three mandarins each, an abacus, ropes of different lengths, a long stretch of fabric, a parachute with twelve segments of different colours, hula-hoops, fans, a whiteboard or flipchart, and music instruments such as hand drums and castanets.

The decoration of the room with props and instruments announces a playful dance and math session filled with colours, shapes and pace. Children and pupils are likely to commenting spontaneously on the numbers spread over the floor, recognising and pronouncing them in different mother tongues, explaining and translating them—triggering an immediate interaction with the math subject. Someone maybe wonders why there are letters on the floor too. An interactive *space of learning* is established before the teacher summons everyone to gathering for the first dance, or the following ethnomathematical introduction.

Ethnomathematical Introduction—Historical and Linguistic Aspects of Mathematics

As the children or pupils recognise and discuss the Roman, Arabic and Indo-Arabic numbers, their engagement calls for a clarifying ethnomathematical run-through. Historically, various cultures and languages have contributed to the knowledge formation of the math subject. Culture and language keep affecting mathematical thinking: in the way numeric systems are named and how numbers are counted in different mother tongues represented in the group. Notions such as *twelve*, *forty-five* or *half past eight* may sound confusing to those still unfamiliar with the majority language. So much easier then, if the child is given an opportunity to translate these notions to their mother tongue, where they may count *'ten-two'*, *'five-and-forty'* or know when to come in time by *'half nine'*.

Throughout, *Dancing Math* links math to language, not least in *meta-talks* after each dance activity, where the math content dealt with in the dances is discussed and the reasoning in turn is translated into formalised math language. This ethnomathematical introduction additionally lowers language thresholds. Keen on counting in Spanish, Portuguese, French, Italian, Rumanian, Hindi, Nepali, Bengali, Urdu, Arabic, Farsi, Igbo, Mandarin and so forth, the children are invited to participate and contribute with their special knowledge and obvious skills in math.

> What are these curlicues lying all over the floor? Numbers! Arabic numbers! Aren't there letters too? Yes, Roman numbers! Where have you seen Roman numbers? On churches! On old clocks. On old houses! In book chapters. The movie Rocky III! How does the British King, Charles the *third* write his royal name? Charles III.

Go through a number scale with Roman numbers from *I* to *XII*, like on an old clock dial. Which letter did the Romans use for the number *one*? An *I*. Write an *I*, with *serifs* for better readability—a word maybe from Dutch *schreef*: *line* or *dash*. What is *two* in Roman numbers? *II*, two I's. *Three*? *III*, three I's. *Four*? Four I's …? Sure enough it would be logical to continue according to the *one-to-one-principle*, one *I* for every *one*, but it's getting hard to read, you almost need to check and count how many I's. Let's skip *four* for a second, what about the Roman number *five*? A *V*. So how is *four* then, and *six*? An *I* before the *V* and an *I* after the *V*: one *less* than *five* and one *more* than *five*! Logical! *Seven* then? *VII*! *Eight? VIII*! *Nine*: *VIIII* …? No, that would be unclear again. Let's skip *nine* and go over to *ten* first. *X*! Now, let's look back at *IV*, *V* and *VI* and then look at *nine*, *ten* and *eleven* … That must be *IX*, *X*, *XI* then. And *twelve* then must be … *XII*!

In the structure of the Roman number system, we see a range of applications of the one-to-one-principle, as each *I* stands for any counted unit. Later on when we dance the clock we'll come back to the Roman numbers *I* to *XII*.

Continue making language connections. Which language did the Romans speak? *Latin!* Does anyone here speak a Latin language? How do you say 100 in your language? *Cien. Cent.* 1000? *Mil. Mille*. Relate the Roman number *C* to *percent—per hundred*, hundred *cent* to a dollar or a euro, a *centimetre*—a hundredth of a metre. Connect *M* to *millennium*, *millimetre* or *per mill—per thousand.*

How did the Romans write *two thousand*? *MM*! The one-to-one-principle again. As in *two thousand and twenty*: *MMXX.* And *two thousand and twenty-two*: *MMXXII.* What about *nineteen hundred*? *MCM*—a hundred less than two thousand*!* *Nineteen hundred nineteen*? *MCMXIX!* Ten less than nineteen hundred and twenty.

For sure it is much handier to write 2000, 2020, 2022, 1900, 1919 and 2022 compared to the Roman way. Each digit's position in a number has a value to be multiplied by 10. When a number like 2022 lacks hundreds we write a 0, *zero* in the hundreds' position.

Who came up with zero? Many students have problems with zero, so it wasn't for nothing a symbol for zero was invented relatively late in math history. So let's check the other group of numbers on the floor, those used in the Arabic world as in Iran and Afghanistan. Ask for help to write *one* in Arabic: ١, واحد, (*waahed*). *Two*: ٢, اثنان, (*ithnaan*), a one with a little curl. *Three*: ٣, ثلاث, (*thalaath*), a one with two curls. Pick up the sheets with ٢ and ٣, hold them up and turn them a quarter turn … What do you see now …? *2* and *3*! That's why we call 'our' numbers *Arabic* numbers. We borrowed them from the Arabs, who in turn borrowed them from …? India, including the symbol for zero. What is *zero* in Arabic, or Farsi? صفر (*sifr*). How is the symbol written? A dot: ٠.

And so it was in India around 500 CE that the mathematician *Aryabhata* defined *zero* and symbolised it by a dot. Ask someone who speaks Hindi, Bengali or Thai what *zero* is in their language: *shoonye*, *shoonya*, *shoono*, *shoon*, which all go back to the Sanskrit *sunyata* which means *void* or *emptiness*. Which was translated into Arabic as *sifr*. And was later passed on to Latin as *zephirum* and on to some languages like English as *zero* and *cypher*, via the French *zéro* and *chiffre* … the very name for *number*! So when we now all speak about *zero* it comes from Arabic *sifr* for *nil* which in turn is the Arabic translation from Sanskrit *sunyata* for *void.*

It was the Persian-Arabic mathematician *al-Khwarizmi* (الخوارزمي) who later on around 800 CE translated Aryabhata's numeric system and advanced mathematics, and did new trailblazing work in mathematics. His name lives on in the word *algorithm* and his book on *hisaab* (حساب) and *al-jabr* (الجبر) gave name to *algebra*. Many children are familiar with *hisaab* as Arabic or Persian for *math*, *counting*, *bill* and *computer*.

Many children are keen on counting in their mother tongue. Handling notions like *twenty-two*, however, can be tricky in the beginning. But if instead we say *two-ten-two*, as in Serbian or Croatian *dvadeset dvije*, it starts making sense. And *thirty-five* may be treacherous if one is used to *fünfunddreißig* in German. French-speaking children for sure must have an arithmetic head who say *quatre-vingt-dix*, *four-twenty-ten* for *ninety*. Danish-speaking children must have the biggest arithmetic heads, who also multiply by twenty and even halve twenties, halfway to the next full twenty, as in *halvtreds* for *fifty*, *tres* for *sixty*, *halvfjerds* for *seventy*, *firs* for *eighty* and *halvfems* for *ninety* …

Connecting language to mathematics is inclusive for children and pupils who speak other mother tongues. Ask someone to teach everybody how to count from one to ten in Japanese, Ukrainian or Greek, and listen if you can hear similarities. It all makes the ties between language and math become clear: between language and mathematical thinking. This way teaching becomes intercultural.

Time for dance.

Arithmetic

In the following, lesson set-ups are sketched with a logical flow of dance activities highlighting essential mathematical notions, featuring the properties and manipulation of numbers, moving on to geometry and concluding with the clock.

Double and Half

Music: free flow, e.g. Omar Faruk Tekbilek: A Call to Prayer.

Everyone spreads out over the floor. Instruct the movement in three tempi.

> Walk freely over the floor, criss-cross. Walk relaxed, light feet, arms hanging down. Change direction when you meet someone else, make eye contact and turn away at a *sharp angle*.
>
> Stop. Stretch arms up, rise on tiptoe. Open arms to the side, spin round, in one direction, in the other direction. Stop, stretch arms up, rise on tiptoe.
>
> Walk around freely in the room again, now to a swifter beat, a springy though relaxed walk, change direction when meeting someone. Quiet feet.
>
> Stop, stretch arms up, rise on tiptoe, spin in one direction, in the other direction.
>
> Same again, now on a run, vigorously, dodge quickly when meeting someone else, faint, spread out your arms and fingers like an insect.

Bring in the notion *double.*

Walk around freely again, like in the beginning, at a calm pace, one by one, covering the whole floor. Change direction at a right angle when you meet someone. When I say *double*, then next time you meet someone you join then and continue two by two, shoulder to shoulder, moving on side by side.

Double!

Everyone joins up in pairs and moves on, crowding the floor. If someone is left over, step in and move along with that dancer.

Stop, stretch arms up, rise on tiptoe, spin, one direction, other direction.

Continue walking over the floor two by two, until I say *double* again. Then when one pair meets another you merge and move on four by four, shoulder to shoulder, maintaining the amenable movement over the floor.

Stop, stretch arms up, on tiptoe, spin.

Next time I say *double* you merge again, four by four become eight, shoulder to shoulder.

And so forth, eight by eight become sixteen by sixteen. The lines grow longer and longer, the movement never ceases.

Don't stop – or you'll get roots under your feet! Everyone finally moves carefully in one big line billowing over the floor, careful not to break the line or stand still or bump into a wall. Before they do, they back a bit, redirect, fumbling and sliding in a keenly coordinated movement.

Introduce a new mathematical notion: *half.*

When I say *half* you split up again and go back to those you walked with the other minute, before the latest merge into the big line. Next time I say *half* you split up again – going backwards in the order of doubling.

Half! The long line splits in the middle. *Half*! The two lines split in two again. And so forth, until everyone is moving shoulder to shoulder four by four, then two by two. As the room for manoeuvre grows bigger the movements grow freer.

How many at the time will you be next time I say *half*? One by one! Then you go disco dancing!

Say *half* one last time, change music.

Music: driving, steady beat, e.g. Omar Faruk Tekbilek: Imaginary Traveller.

Everyone dances freely, by themselves, at a faster beat, disco or techno. Dance like Amanda! Everyone imitates that dancer. Dance like Arkan. Everyone moves like that dancer, and so forth, through the whole group.

Turn off the music.

Stretch arms up, rise on tiptoe, lower arms to the sides, turn to one direction, to the other direction. Sit down, take a breather.

Meta-talk on Double and Half

Discuss the dance activity. Help put words to the movements and everything that unfolded in the room. Spur the use of appropriate dance language: we walked one by one, changed directions, at sharp angles, rose on tiptoes, stretched our arms straight up, turned, moved along with someone we met, continued two by two, shoulder to shoulder, gliding over the floor, then four by four, and eventually we moved together in one long billowing line.

Go into the notion *double*. Support the correct use of language for math.
How can we explain and define *double*? In the dance you proved you are aware of what double is. How can you explain this concept with words in a clear way?
Someone may say 'the same once more'. Someone else says 'equally many again', 'twice the same', 'times two', 'one plus one' or 'the same amount again'.

Take notice of all suggestions and modes of explanation, help out with distinctions in math and in language. Elaborate on two alternative models that are likely to crystallise, *addition of the same number* or *amount*, and *multiplication by two*. Sort them out together.

What does *same* mean, and *more*? Equally many: the same number or amount is *added*. What is the operation *to add* called in math language? *Addition*. What is the sign for it? The plus sign.

Refer back to the dance, how we started one by one and then you said 'double'. Ask one dancer to stand up. Here we have *one*. You say that for *double* we need the same amount or equally many again. That means *one more*. Ask *one more* to stand up. What do we do with them? Put them together. How many are they now? Two.

Formalising on the Board

With the dancers standing on the floor, write down 1 1 2, or ask someone to keep the record on the board. Where do we put the plus sign? Between the two 1:s. What is the sign for the result, the *sum*? The *equals sign*. Write the addition $1 + 1 = 2$. Read the equation loud all together: *One added to one equals two*. Repeat the doubling, from two to four. We have our two dancers here. For double we need *two more* or *another two*. Ask *another two* dancers to get up, put them together, write the addition on the board. Continue as far as we went in the dance. List on the board the equations under each other: $1 + 1 = 2$, $2 + 2 = 4$, $4 + 4 = 8$ and $8 + 8 = 16$. On the floor we have up to eight or sixteen dancers.

In the other way of explaining maybe someone said: 'two times the same'. Which operation are we into when we speak of *times*? *Multiplication*. How many *times* is double? *Two times*. We started moving around one by one. Ask someone to stand up. Here we've got *one time* a dancer. How many times do we need to get double? Two times, or twice. Ask someone else to get up. That's another single set of one. We put them together. *Put together* … it does remind us of adding, as we do increase with the same amount: *one more time*. We will come back to the playful logic of mathematics and the relation between multiplication and addition. How many are they now? *Two!* The same result as in the addition of the same amount—two different ways of thinking the notion of *double*.

Write on the board, beside the corresponding addition, 1 2 2: We've got *one*, *two times*, which makes *two*. What is the sign for multiplication? × or •. What is the sign before the result again, in this case the *product*? The *equals sign*. Write the multiplication $1 \times 2 = 2$. Read the equation aloud together: *One multiplied by two equals two*. Repeat doubling, from two to four. Now we've got two, *one time* or *once*. For double we need two times or twice, that is one more time. Ask two pupils to rise, so we get two times two pupils. We put them together. Write the multiplication by two on the board, beside the corresponding addition. Continue as far as we got in the dance. On the board we now have, under each other and beside the corresponding additions, the eqs. $1 \times 2 = 2$, $2 \times 2 = 4$, $4 \times 2 = 8$ and $8 \times 2 = 16$. Just like the arrangements of dancers on both sides of the room.

Summarise the two ways of reasoning about *double* and how we arrive at the same result, from one to two, from two to four and so forth—*addition of the same amount* or *multiplication by two*.

Every number can be doubled: *one*, *two*, *three*, *four*, *five* and so on. Arrange dancers in the room to make the concept concrete.

Go through *half*. Someone may say 'divide equally', someone else 'divide in two' or 'split in the middle'. Go to the arrangements on the floor. Here we've got eight pupils, who can be divided in *halves*, just like you did in the dance: you divided the dancers into two parts with equal numbers in each, two *halves* of four. Then we *halved* the groups again, on to two or pairs, and then we halved these ones again and so we were back to one by one and then we had the disco dance.

Can you divide *one* into halves? Preferably not, a shame for such great dancers. Everyone knows you can split an apple into two halves, and you can even halve the halves, into *quarters* or *fourths*, which in turn can be halved into *eighths*, and so on. But today we stick to whole apples and whole dancers. Go through a line of eight pupils. *One* we can't halve, not today. Could we halve *two*? So we did. Can you halve *three*? Again, not if we want to keep everyone whole, pity on the one in the middle. *Four*? So we did dancing. *Five*? Then some poor buddy comes to grief again, but we agreed to stick to whole dancers today, whole apples, whole numbers. Six? Seven? Eight? It looks like every other number can be divided into two, every other number not …

Let's step it.

Even and Odd Numbers

> Everyone moves freely over the floor and stamps an equal and regular pulse. When the rhythm is clearly established, accentuate every other stamp, counting high: one *two* one *two* one *two* … Keep in step!

Make sure everyone joins into the common rhythm and pulse. Ask what the time is called in music terms when you count *one two one two one two* …? *Duple time*!

Now let's count all together from *zero* to *twenty-four* while we stamp and stress every other foot. We start at *zero* on the first hard stamp, loud and clear. First, let me hear the beat of the feet: one *two* one *two*. And where do we start from? Zero! Ready? ... *0* 1 *2* 3 *4* 5 *6* 7 *8* 9 *10* 11 *12* 13 *14* 15 *16* 17 *18* 19 *20* 21 *22*.

Now let's do it again. We stamp the duple time and count aloud from zero to twenty-two, but this time we only say those numbers aloud we stamp on hard, the other numbers we just whisper or count quietly in our heads. What's the rhythm of the feet? What do we start on? Ready, steady ... *0* (one) *2* (three) *4* (five) *6* (seven) *8* (nine) *10* (eleven) *12* (thirteen) *14* (fifteen) *16* (seventeen) *18* (nineteen) *20* (twenty-one) *22*. Good!

Do it once more and ask someone to write the accentuated numbers in a column on the board.

Now let's stamp and count again, but now we count those numbers aloud we stamp softly. A bit of a coordination exercise. Tip: 'stage whisper' the unstressed numbers.

Ask the pupil by the board to write down the new numbers in a column to the right of the first one.

All ready? Let me hear you stamp the rhythm ... one *two* one *two* one *two*. What do we start on ...? Ready, steady ... (whisper: zero!) *1* (whisper: two) *3* (four) *5* (six) *7* (eight) *9* (ten) *11* (twelve) *13* (fourteen) *15* (sixteen) *17* (eighteen) *19* (twenty) *21* (twenty-two)! Brilliant!

Now check the first column, with the numbers we stamped on hard. Let's skip the zero for now: 2, 4, 6, 8 ... 18, 20, 22. Can you halve these numbers, that is, divide them by two? Sure, so we did. What do we call these numbers? *Even* numbers. Check the second column, with the numbers we did not accentuate: 1, 3, 5, 7 ... 17, 19, 21. Can you halve those? Not if we stick to whole numbers. What do we call these numbers? *Odd* numbers.

Two Times Table

Look at the column with the even numbers, the table we get by stamping the duple time and counting from 0 till 22. What is the table called? The two times table! Discuss what a multiplication table is. In the two times table all numbers are multiplied by two, one after the other, from 1 to 11. On top is the zero, let's wait with that one. It did get into math late in history.

Write the multiplications by 2 left to each even number in the column, in formalised math language: $1 \times 2 = 2$, $2 \times 2 = 4$, $3 \times 2 = 6$ ... $10 \times 2 = 20$, $11 \times 2 = 22$. Point particularly at the multiplication of 10 and 11 by 2, giving the products 20 and 22, an aesthetic aspect easy to store in our graphic memory.

Point out that the table reveals a *repeated addition of 2*. That's what happened when we stamped and counted aloud: for each new hard stamp we counted two more numbers. Remind them about the discussions on the double of two, which we could explain as *one more time*. Worth keeping in mind, however, is that a *multiplication table* first and foremost is just what it is called—all numbers *multiplied* by a certain number.

Three Times Table

> If we get at the two times table by walking around in duple time, stamping hard on every other foot, how then does the three times table sound in the feet …? One two *three* one two *three* one two *three*: a walk in *triple time,* the accent on every third stamp.

Establish the joint pulse. Count 1 2 *3*, 1 2 *3*, 1 2 *3* … In earlier times, children played train games, speeding up the rhythm, stamping and doing circular movements with the arms besides the body, upper and lower arms at a right angle. Nowadays these associations with steam locomotives are no longer common, but then again the numbers of steam train enthusiasts of all ages are increasing all over the world.

> How do we do the three times table our way? We stamp the triple time. What do we start on? Zero! How long do we count? Until thirty-three! Ready? Stamp … count loud and clear: *0* 1 2 *3* 4 5 *6* 7 8 *9* 10 11 *12* 13 14 *15* 16 17 *18* 19 20 *21* 22 23 *24* 25 26 *27* 28 29 *30* 31 32 *33*. Now as the triple time 'sits' firmly in our feet we stamp and count again from 0 to 36 and scan the numbers aloud on which we stamp hard, the other ones we whisper or count quietly. Ready, steady … 0 ·· 3 ·· 6 ·· 9 ·· 12 ·· 15 ·· 18 ·· 21 ·· 24 ·· 27 ·· 30 ·· 33.

Do it again while someone writes the accentuated numbers in a column on the board, and then completes with the formal equations $1 \times 3 = 3$, $2 \times 3 = 6$, all the way to $10 \times 3 = 30$ and $11 \times 3 = 33$. Now we've got the three times table.

When triple time sits firmly in the legs, dance the waltz on the table of three, either one by one or in pairs, waltzing away over the floor. Accompany the dance on the piano or with a drum, or let someone play the rhythm on the drum or the guitar. You can also dance the waltz and sing the three times table:

> Zero times *three* that makes *ze*ro, yes *ze*ro …,
> and *one* times a *three* that makes *three*, that makes *three* …,
> and *two* times a *three* that makes *six*, that makes *six* …,
> and *three* times a *three* that makes *nine* – oh yes *nine!*
>
> *Four* times a *three* that makes *twelve*, that makes twelve …,
> and *five* times a *three* that makes *fif*teen, yes *fif*teen,
> and *six* times a *three* that makes *eigh*teen, yes *eigh*teen,
> and *seven* times a *three* that makes twenty-*one*, oh yes twenty-*one!*
>
> *Eight* times a *three* makes twenty-*four*, yes twenty-*four* …,
> and *nine* times a *three* makes twenty-*se*ven, yes twenty-*se*ven …,
> and *ten* times a *three* that makes *thir*ty, yes *thir*ty …,
> and *eleven* times a *three* that makes thirty-*three*, oh yes thirty-*three!*

Note that everyday language is used in the song, while appropriate math language is: 'Zero multiplied by three equals zero', and so on.

Note also how that tricky issue sneaked its way in: *multiplication by zero*. It tops all the tables on the board, as we started all the rhythms stamping on the zero. Let's make note of it and come back to it in a little while. First we finish off the remaining tables, beating the times with our feet. Observe once again how easy multiplying 11 by any number is. In the 2 times table we get $11 \times 2 = 22$, in the 3 times table $11 \times 3 = 33$. What about the 4 times table? 44! 5 times? 55! 8 times? 9 times?

Four Times Table and On

How does the four times table sound in the feet? Must be quadruple time! One two three *four* one two three *four* one two three *four*. What do we start on? Zero! What do we end on? Forty-four! Stamp and count … ready, steady … *0* 1 2 3 *4* 5 6 7 *8* … 39 *40* 41 42 43 *44*. Again, now count aloud the numbers we stamp on hard, whisper the other numbers: *0* ··· *4* ··· *8* ··· *12* … *40* ··· *44*.

Ask someone to record this in a table on the board and fill in all the equations in front of the four times numbers: 1 × 4 = 4, 2 × 4 = 8 … 10 × 4 = 40, 11 × 4 = 44.

What does the five times table sound like in the feet? Quintuple time, of course, stressing every fifth stamp: 1 2 3 4 *5*, 1 2 3 4 *5* …

Stamp and count the five times table from 0 to 55. Ask someone to keep record on the board and write the equations of the whole table.

Continue with the six times table, and so on, until the ten times table.

How do our feet stamp the different tables: two times, three times, four, five and so on? In the even numbers' multiplication tables – that is on the even times – the same foot stamps hard throughout the whole table, in the odd numbers' tables – the uneven beats – both feet alternate.

The logic and regularity in mathematics is manifested clearly in the body!

Multiplication by Zero

Zero is a tricky thing for many. As mentioned in the ethnomathematical introduction, it took a long time in math history before zero was 'invented' to give name to what is 'null and nothing', in number or length or weight.

Multiplication of a number by zero is even more knotty to figure out. But thanks to our way of dancing rhythms and stamping the multiplication tables we've actually solved the problem. The first stamp in each table starts precisely on *zero*. Thanks to this, the emphasised stamps in, say, the three times table, give us the numbers 0, 3, 6, 9 and so on until 30 and 33. And in that rhythmical logic 0 multiplied by 3 obviously is … 0! The same when we dance the two times table: 0 multiplied by 2 equals 0. Or the four or five times: 0 × 4 = 0, 0 × 5 = 0. And so on. Any number multiplied by 0 gives us the product 0.

We can also reason in the logic of math by reminding ourselves that multiplication unfolds as repeated addition—by 2, 3, 4 etc. If we then backtrack in the table from eleven times a number to ten times and so on, we observe a repeated subtraction going slowly and steadily towards … zero.

One way to sensualise and concretise this mathematical thinking is by means of eleven paper saucers or plates with three mandarins on each. Excellent for a fruit break after all that stamping and counting. You can also put pea or kernel bags or small balls on the plates or whatever is at hand. Run through the three times table again and put one plate with three mandarins at the time on the floor.

Let's rehearse the table and check on the board. I put down *one* plate with three mandarins, how many mandarins does that make? *Three*. I put down a *second* plate with three mandarins I've got ... *six* mandarins. *Three* plates with three mandarins each makes ... *nine*, and so forth. *Ten* plates with three mandarins makes a total of *thirty* mandarins and finally *eleven* plates with three mandarins makes altogether *thirty-three* mandarins.

Now remove one plate at the time and repeat the three times table backwards. Relate to the table written out on the board.

We start from eleven plates with three mandarins, totaling thirty-three mandarins. I take away the eleventh plate, how many plates with three mandarins are left then? *Ten*. How many mandarins does that make? *Thirty*.

Continue removing one plate at the time and count down together.

As I take away the fourth plate with three mandarins, how many plates with three mandarins are there left? *Three*. How many mandarins does that make? *Nine*. We take away one more plate, how many plates are left? *Two*. How many mandarins? *Six*. We take away the last but one plate with three mandarins, how many plates do we have left? *One* plate with three mandarins. How many mandarins is that? *Three*. So when we now take away the last plate with three mandarins, how many plates with three mandarins are there left? None: *zero ... Zero* plates with three mandarins. How many mandarins does that make...? *Zero! 0 multiplied by 3 equals 0.*

If there were five pears on each plate, how many pears is zero plates with five pears? Zero: 0 multiplied by 5 equals 0. If the annual soccer football tournament is coming up and all eleven players in the school team got a stomach illness so our school won't have any team of eleven at the tournament, then how many students from our school will participate? Not one, or none: 0 multiplied by 11 equals 0. How much is zero multiplied by 100? 0! By 1000? 0! Zero multiplied by three hundred and fifty-four billion nine hundred and sixty-five million eight hundred and seventy-two thousand nine hundred and twenty-four makes ...? Save your breath teacher: zero! Any number multiplied by zero equals zero.

From Zero to Infinity

However. That previous number sure was a big one. But imagine a much bigger number, and a still bigger one. Imagine you keep counting your whole life and then your children continue and their children and their grandchildren, and so on in all *eternity* ... Just like mathematicians in history invented a symbol for smaller and lesser things down to *nothing*, they also invented a symbol for a number bigger than any number you can think of or count to. Whenever you try to think of an enormously big number, you understand that you could continue counting, in all ... *infinity*. What is that 'number' called? *Infinite* or *endless*. What is the symbol? ∞, the *infinity* symbol. The number obviously does not exist, it is an *unreal* number. It is *imaginary*; we can *imagine* it, never reach it, no matter how long we keep counting.

Just like *zero*. *Innumerable* or *countless* is an *imaginary* amount. *Immeasurable* or *endless* is an *imaginary* length, a line you can draw out into space, it continues and continues straight through and beyond billions of galaxies. Imagining endlessly big numbers or endlessly long lines as well as endlessly small amounts or short lengths is what mathematicians and philosophers throughout times have pondered over, like artists and children in wonder. Who hasn't figured doubling two, four, eight, sixteen, thirty-two, sixty-four, one hundred and twenty-eight, and on and on and on. Or thought of halving an apple in two, four, eight, sixteen and on and on into smaller and smaller parts—at which point is there nothing left ...?

So it is argued in infinitesimal mathematics—math of extremely small quantities—that if you multiply a number by 0, and let the number increase and increase towards ∞, then the product of 0 times that number finally equals ...1! Puzzled? Later on, when we dance geometry and deal with all the polygons, and take the opportunity to clear out all fractions, we will return to the mathematical thinking about infinitely small and big numbers. And come back to this riddle about the multiplication by zero of a number tending to infinity. Until then, keep philosophising if no matter what it just has to be zero. We'll see.

As Many, More, Most—Few, Fewer, Fewest

The group divides into two rows along the shorter walls of the room. One group counts aloud how many there are in the row opposite. Help by pointing at every head during the counting.

Repeat how many dancers there are on each side—for example 11 and 13. Are there as many in both rows? Obviously not, you clearly hear and see they are unequal in number. Which row is more? The *13*. If we weren't here in the room and able to dance and see everything in the concrete but instead just *think* the numbers 11 and 13 in the abstract, how do we determine which number is the bigger? What determines whether one number is bigger than another? The *number line* or *number scale* indicates a fixed *number order*—1, 2, 3... and so on—just like we counted aloud. In our example 11 comes first in the order, then comes 12 and then 13.

Some may have remarked that there are two *more* in the bigger row. If there are two *more* in the row of 13, how do you say the opposite in the other row? There are two *fewer*. They are the *fewest*, those on the other side are the *most*.

So how do we proceed to even it out so there are as many on each side? One suggestion may come up that someone should change sides. Why does that solve our problem? First, you count 11, as many as on the other side. Then you see that the rows *differ* by two, or the *difference* is two, and you divide the two equally. The surplus or difference is *halved*. We will come back to *subtraction* in the next dance.

One changes sides, now there are 12, as many on one side as on the other side.

Ask for suggestions on how to divide a larger group into two *equally big* groups: how to divide a group in two, in two *halves*, how to *halve* a group. Common suggestions are: everyone takes someone else's hands and then every pair splits up, one goes to one side, the other one to the other side; everybody lines up in two lines in the middle of the room so that each person stands opposite to someone else, similar to the first suggestion to make pairs two by two and then split up; all gather in one bunch in the middle of the room and send every other one over to one side and every other one to the other side of the room, like in a sports class when setting up two teams; give everyone in turn a number, one, two, one, two, then the ones go to one side, the twos to the other; count everyone, 1, 2, 3, 4 ..., then the even numbers go to one side, the odd numbers there. Once a suggestion came that all the boys went to one side, and all the girls to the other side, then both gangs split up in two, so you get two groups with as many girls as guys ... All suggestions draw on the condition that the group can be halved, that is if it is even in number. If not, you always have a dance teacher up your sleeve.

How do we divide a group into *three* equally big parts? One suggestion is for everyone to put themselves in groups of three and then split up. Or everyone gets a number, one, two, three, one, two, three and so forth. Continue working out division by four, five and so on. Later on we will return on how to shape two lines, one line with twice as many as in the other, or three times as many, or four and so on.

Pals of 10—Parts and Whole, Addition and Subtraction

Music: rhythmical duple or quadruple time, e.g. Radio Tarifa: Rumba Argelina; see detailed description of the basic steps in 'Galloping or Skipping Two by Two' in Chap. 2.

Now we have arranged two rows with equal numbers of dancers at each short end of the room, standing beside each other, everyone facing a peer dancer on the opposite side. Join one row and demonstrate the steps with a dancer standing opposite, show it twice then they return to their row.

> Gallop towards each other, hands on the backs, meet in the middle, gallop a few rounds around each other, change sides.
>
> Everyone rehearses to the music two by two. Those standing clap the rhythm cheering each dancing pair. If the group is big they dance two pairs at a time, starting at either end of each row then moving inwards.
>
> When everyone has danced two by two all gallop forward at the same time, meet in the middle, lift their arms, wink and shout 'bye, bye!' and gallop backwards, hands on backs, a few times back and forth.

Then only five on each side keep standing beside each other, the others sit down in a bunch by the side of their row. The five standing will dance, those sitting will be their side's *counting masters*.

> How many dancers are standing on each side? Five. How many dancers are they all together? Ten. This is the *pals-of-ten* dance.

Write or ask someone to write the numbers 5 and 5 on the board with some space in between, encircle them, and write the addition 5 + 5 = 10 in between. In a while, the counting masters on the one side will determine how many of the five dancers will gallop to the opposite side and gallop back with as many dancers from the other side.

Go in and show an example. Let's say they decided *one* dances forward: Gallop on to the dancer opposite, gallop backwards while the peer dancer gallops along to your side. Thank them and ask them to return to their side.

Now the counting masters on the first side determine how many will dance forward and bring as many back. They whisper, agree on a number and pass the verdict—it can take time—*three*! Put on the music.

Three gallop to the other side and bring back three from there. All clap the rhythm. Pause the music. Check the regroupings in the room together. Start with those who went to fetch the ones from the other side.

How many were you at the beginning? Five. How many joined? Three. Did you get more or fewer? More. Plus or minus? Plus. What is the arithmetic operation called when something is added? *Addition*. How many equals three added to five? Eight, the *sum*.

Write or ask someone to write the addition on the board: 5 + 3 = 8.

What is the situation on the other side? How many dancers were you at the beginning? Five as well. How many galloped away? Three. Did you become more or fewer? Fewer. Plus or minus? Minus. What is that operation called? *Subtraction*. How many equals five subtracted by three? Two, the *difference*.

Write the subtraction on the board: 5–3 = 2. Read it aloud together: Five subtracted by three equals two.

Now there are 8 on one side, 2 on the other side. How many does that make together? Ten! No one fell out, no one jumped out through the window, no one died on the way. Ten were at the beginning, 5 and 5; ten are still there now, 8 and 2: 8 and 2 are *pals of ten*.

Review on the board: write 8 and 2, encircle them and write the eq. 8 + 2 = 10 in between.

Now the counting masters on the other side get to decide how many of the two remaining will dance to the opposite side and bring back as many—*two*! Put on the music, they perform the dance. Check the regroupings. Start with those who increased: 2 + 2 = 4. Check the other side, who decreased, 8–2 = 6. Let them count and solve the new sum and the new difference, write it all on the board. Encircle 6 and 4, which also add up to 10 as they are *pals of ten*, write 6 + 4 = 10.

Continue and let them dance and explore other combinations. Eventually all variations of pals of 10 are reviewed: 7 and 3 as well as 3 and 7, or 0 and 10 as well as 10 and 0. Ring in all pals of 10, write out all occurring additions and subtractions, emphasise all additions sum up to 10—showing all parts of a whole 10.

Helping Tools

An abacus with ten beads a row reinforces visually the pals of 10 that were revealed in the dance: the starting position 5 and 5, 8 and 2, 6 and 4, 10 and 0, etc. Another aesthetic experience of this fundamental arithmetic notion, visual and tactile, is to make ten big balls in art class, in papier-mâché or other material can, thread them on a rope or a wire stretched between two walls, and display different pals of 10 every day.

Pals of 20

This dance is a natural progression of pals of 10, with ten dancers on each side, totalling 20. Dancing pals of 20 entails dealing with transfers of ten, which is elaborated on later. For now, the kinaesthetic experience suffices to spatially sense how pals of 20 are composed, along with the graphic formalisation in equations on the board, as a first mentioning of the transfer of ten.

Galloping Exam on Double

Music: Duple or quadruple time, paso doble, circus music; helping tools: long fabric, hand drum, castanets; see detailed description in 'Galloping and Jumping Over the River' in Chap. 2.

Everyone lines up behind each other at one side of the room. Hands on hips. A long piece of fabric lies across the floor in the middle. One by one the dancers gallop forward, jump over the fabric and build a new line from the opposite wall. They wait for their turn, when the previous one jumps over the fabric they start off galloping.

If you think the leaps after the first round could be a bit higher, lie down on your back on the fabric so they get to jump over you, or put a cushion, a box or a bar in the middle.

When everyone has jumped one by one two rounds, they line up again as from the start. Remind them of the notion *double* or *multiplication by two.*

First they gallop one by one again. They jump over the fabric in the middle, gallop on, turn in one direction in a wide arch, gallop back along the side wall, jump over the fabric again and come back to where they started. Every alternate dancer turns back to the left side, every other one to the right.

Two lines are shaped opposite each other by the shortening line of dancers waiting to set off. When everyone has galloped one by one, one from each line takes the other by the hand, their outer hands on their hips, and they gallop two by two. The pairs gallop forward, jump, gallop on, split up and gallop back one by one, each in the same direction as before. On the way back they jump over the fabric and join their lines again in the back. When they've all galloped and jumped two by two, they dance four by four, two from each line, gallop, jump, gallop on, split up two by two and turn back in the same direction as before, jump over the fabric again, and build on the two lines. Then four from each line join and gallop eight by eight, then sixteen by sixteen, and at last all together in a grand finale, galloping and jumping like one single examinee at a run to the highest grade.

As Long, Longer, Shorter, Twice as Long, Twice as Many

Everyone lines up in two straight parallel lines next to each other.

Make sure the lines are *equally long* and that there are *equally many* in both lines. How do you work that out? Everyone positions themselves so they face someone in the opposite line. Easiest if everyone holds each other's hands in pairs. Go in yourself if the group is uneven in number.

Just like we did earlier, the group has now divided into *two halves*.

How do you proceed so there are *double as many* or *twice as many* in the one line as in the other? Meaning of course there are *half as many* in the other line.

Some may suggest that every other person, or every *second* person in the one line goes over to the other line. Does this solve the problem? It looks as if the new line becomes three times as many, not double—good to keep in mind for later on.

They position themselves so each one in one line faces two in the opposite line. This means the group is *divided in three* and that *one third* stands in one line and *two thirds* in the other line.

This requires the whole group to be divisible by three for everyone to be included. If not, there is a *remainder* when the division by three does not go evenly. Then the *remainders* can just stand by and keep an eye so everything is correct.

How do you proceed now so there are *three times as many* in one line as in the other? Obviously each one in the thinner line must have three people opposite. Hold each other's hands, one in the one line, three in the opposite line. Thus the group is divided into four, *one fourth* standing in one line and *three fourths* in the other line.

This leads to the same result as in the earlier reasoning about double as many, where every second person in the one line of two with equally many went over to the other line: the result is a line with three times as many, not double: three halves of the total in that line and one half staying in the other line.

By now we have broken the code so we can continue dividing the group so we get five, six, seven times as many in one line as in the other line.

Finally, go back to the two lines where there are *double as many* in the one line as in the other.

Can you make the line with *half as many* dancers *double as long* as the other one …? In other words, can you make the line with *double as many* dancers *half as long* as the other line?

Let them reason towards a solution. Confirm the difference between *length* and *amount*. In principle three or even two people can make a line double or three or even ten times as long as a line with double or three or ten times as many people.

Galloping and Stamping in Two Equal Lines

Music: rhythmical quadruple time, e.g. Juan Peña Lebrijano & Orquesta Andalusí de Tanger: Las Mil y Una Noches.

Everyone lines up in two lines with as many in each line again. Show the dance steps:

> Gallop four times with one leg in front, change legs, gallop four times with the other leg in front; change legs again, two gallop steps with the first leg in front, change legs, two gallop steps with the other leg in front; change legs again, then four times with alternating legs in front. Instruct with the help of your voice and the rhythm, stressing the first beat: and a *1* 2 3 4, a *1* 2 3 4, a *1* 2, a *1* 2, a *1* a *1* a *1* a *1*, again: and a *1* 2 3 4 …
>
> Both hands rest on the same hip as the leg in front, elbows sticking out a bit forward. At each leg change hands and arms do a wide arc up in the air from the one hip to the other. When the changes come in quick succession at the end on the beats *1* and *1* and *1* and *1*, the hands tap on the beat quickly alternating from hip to hip.

Rehearse the dance steps galloping on the phrase 4 – 4 – 2 – 2 – 1 – 1 – 1 – 1 freely across the floor. Show the dance pattern in the room. The two lines stand parallel besides each other by the one end of the room.

> The first pair, one from each line, gallops forward on the *1* 2 3 4, and a *1* 2 3 4, split up at the other end of the room, one dancer turns in one direction, the other one in the other direction, they gallop back on the *1* 2, and a *1* 2, and *1* and *1* and *1* and *1*, landing back by the parallel lines, facing each other and shaping the beginning of a gateway for the following pairs to gallop through. As the first pair lands there, keeping the pace, the next pair dances forward and curves back, each dancer to their own side, landing besides those in the first pair, building on the gateway for the next pair to start galloping through.

When the last pair has danced through the growing gateway and back, finishing the gateway, everyone is standing in two lines opposite each other again. Keeping the pace, everyone stamps on the spot. Join in and instruct singing and stamping the rhythm:

> One foot forward stamps: and a *1* 2 3 4 – change foot, with a little hop on the upbeat *and a* *1* 2 3 4 – change foot – and a *1* 2 – change foot – and a *1* 2 – change foot uninterrupted for four times – and *1* and *1* and *1* and *1*. Once more: and a *1* 2 3 4 and so on. The beat goes on, everybody stamps.
>
> In the next round everybody claps their hands on the same beat as the stamps, as well as on each upbeat 'and a' marking a foot change.

The rhythm may seem complicated in writing, but like so much else in dance it's most often much easier to do than to say, as is in math danced this way.

The Base-10 Positional System

Three colourful hula hoops lie on the floor. Ask for a three-digit number; for example, 235. Put on the music. Everyone dances all over the floor. When the music stops they compose the number *two hundred and thirty-five* in the hula hoops: *two* hop into the ring to the left, *three* into the middle ring and *five* into the ring to the right. Ask for another three-digit number, turn the music on, everyone dances again until the music stops and they make the new number in the rings. Continue the dance stop game a few more times. Return to the number 235 at the end.

Those who did not get into the rings go and stand in front of them so they can read the number from left to right. Emphasise how the number sounds in speech, pointing at the dancers in each ring: *two hundred and thirty-five*.

Focus on the five to the right. How much is the value of each of these five in their position to the far right? *One!* They are *units*. Together their value is ... *five*, just like it sounds in the number: two hundred and thirty-*five*. Mark the unit position with a plastic coated note *1* in front of the ring.

Let's look at the next ring in the number two hundred and *thirty*-five. What is the value of each of these three in their position? *Ten!* They are *tens*. How much is the value of each of these tens compared to the *units*? Ten times their value. What is the value of the three tens together? *Thirty*, like it sounds in the number: two hundred and *thirty*-five. Put a coated note *10* in front of the hula hoop.

Let's move on to the final ring in our number *two hundred* and thirty-five ... What is the value of each of these two in this position? *One hundred!* They are *hundreds*. How many times is each of them worth compared to the tens? *Ten times!* How many times is each of them worth compared to the units? Ten times ten, a *hundred times!* How much are they worth together? *Two hundred*, like it sounds in the number: *two hundred* and thirty-five. Put a plastic coated note *100* in front of the left hula hoop.

If we lay a new ring to the left of the 100-ring, which number position will that be? *Thousands*. Put a plastic coated note *1000* to the left of the 100-ring. So it goes on, another zero is added each time, a new leap of multiplication by 10 in the base-10 number system.

Put a plastic coated period sign to the right of the units, indicate an imagined row of hula hoops to the right for tenths, hundredths, thousandths. Relate to *percent*, *per mill* or *per thousand* and the Roman figures C and M. Confirm the period sign as a *decimal point* or *decimal sign*, which *decimates* the units.

Addition and Subtraction with Transfers of Ten

Music. Everyone dances freely. Pause the music, say a new number with a zero in it: for example, 205. The requisite numbers of dancers jump in the hula hoops to compose the number. Ask how many people there are in the ring of tens. *Not one!*

No tens. Discuss how handy the number zero is in the positional system, compared with the older number systems, like the Roman or Egyptian, where writing big numbers is clumsy.

Lay another two rows of three hula hoops above the first row. Put a rope between these two rows and the first row, and a plus sign above the far left end of the rope.

Ask for two three-digit numbers. Ask someone to write the numbers in a vertical column model on the board: for example, *423* and *124*. Put the music on, everyone dances. When the music stops nine dancers fill in the first number in the top hula hoop row and seven dancers fill in the second number in the rings below, above the addition line.

Let's add the numbers. How is an addition of two multi-digit numbers done? We must not mix apples and pears, as each position has a different value. Units are added to units, tens to tens and hundreds to hundreds. Where do we start? To the right, by the units.

Ask the *three* in the unit ring on top and the *four* in the ring below to go down and gather all *seven* in the unit ring under the rope. Do the same with the *two* plus *two* tens and the *four* plus *one* hundreds, adding up to *four* tens and *five* hundreds in their rings under the rope. Ask the pupil by the board to keep record of the numbers under the addition line. The sum is *423 + 124 = 547.*

Ask for two new three-digit numbers, with units summing up to more than 10; for example, *215* and *106*. Ask the pupil by the board to write the numbers in the vertical column model. Music. Dance stop.

The two numbers are composed in the rings, as done before. Do the addition of the units. Ask the two groups of *five* and *six* to gather in the *units*-ring under the line, now comprising eleven. Can the record keeper by the board write the sum *11* in the tens column? No, there is only room for one digit from 0 to 9 in each position. From 10 we have added up to a *ten*, which equals a 1 in the next position: a *transfer of ten*. Ask ten of the eleven pupils to step out of the unit-ring and gather a bit to the side. What is their total value? *10!* As many as the one ten in the next ring, the next position. Let the ten units appoint an envoy to the reign of tens. They wave good-bye to their ambassador who walks over—*transfers* or *regroups*—and takes a position above the top hula hoop of the tens. The pupil by the board writes a 1 above the tens. Now all tens add up and gather in the ring under the rope, the one in 215 and no one in 106, and the representative for the ten units slipping in. Their sum equals 2, which makes twenty. Then the 100 s are added, 2 and 1. The whole sum is *215 + 106 = 321*, on the floor as well as on the board or in the minds of keen ready reckoners.

A similar arrangement with hula hoops can be set up for subtractions; for example, 245−126. In the *minuend* 245, one of the 4 tens transfers or regroups to the 5 units—decreasing the 4 tens to 3 and increasing the 5 in the unit ring to 15. This allows for the 6 in the *subtrahend* 126 to be subtracted from 15, giving the difference of 9 units. Then the 2 tens of the subtrahend 126 are subtracted from the 3 tens in the 'new' minuend 235, and the 1 hundred in 126 from the 2 hundreds in 235. The difference between 245 and 126 gives 119.

Geometry

What is geometry? Forms! Shapes! Figures! Bodies in space! Math you can draw! Can you give examples? A circle! A square! A triangle! A ball! A sphere! A cylinder! A trapezium! A line!

Point out the elements of geometry: the point, the line and the plane. All shapes and bodies are composed out of these building blocks of geometry.

What is the difference between a ball and a circle? Both are round, but a circle is *flat* and can be drawn on a paper, while a ball or globe is *spherical* or *spatial*.

Discuss *two-dimensional* and *three-dimensional*. The two-dimensional goes from one point straight out in two directions and forms a *plane*, such as a square or a circle. The three-dimensional goes out in three directions and forms a *space*, such as a cube—a square with a height—or a cone—a circle with a pointed hat. Then what is *one-dimensional*, which only goes out in one direction? *A line!* Is there something like *zero-dimensional*, which does not go in any direction? *A point!* It has neither length, width or height, it is just a *position*, a certain place. Then there is also the *four-dimensional*, adding the *time-dimension*, used to describe the *space-time*.

Figure Dance

Music: driving, steady rhythm, e.g. Omar Faruk Tekbilek: Imaginary Traveller.

Everybody dances freely, crisscrossing over the floor, disco dancing. Encourage variation of movement, arms and legs, spins and leaps.

Say *circle*! Dancing on, everyone moves into a circle in the middle of the room. When everyone has found their place in the circle and are happy with the figure, it dissolves and everyone goes back to dancing all over again. Call out another figure—*square*, *triangle*, *rectangle*, *pentagon*, *parallelogram*, *trapezium*. The figures are shaped on the floor and dissolve in general dance again.

The Circle—for Younger Children

Lay out hula hoops, one for each child to go and stand nearby. Discuss the form of the hula hoops, discuss rings, *round forms*. A round form or figure is called a *circle* in math language. Pick up a round tambourine. Is this drum *round*? It is shaped like a *circle* too. Shake the tambourine, count to three, beat on the three:

On the count of *three* you hop *into the circle*. Are you ready? One ... two... *three*! Yes! On the next count of three you hop *out* of the circle again. One ... two ... *three*! Great!

Repeat a couple of times. Move on to varying the hops. First a hop *forward* into the circle and on the next beat on the three a hop *forward* out again *to the opposite side of the circle*. Then a hop *forward* into the circle and out again *backwards*. Then a hop *forward* in, a hop in the air *half a turn around* and *forward* out again. And so on. Conclude:

> On the next count of three, hop *into* the circle and sit down. Let's try it! One ... two ... *three*! Great! And on the next count of three jump *up* and *out* of the circle again. Ready? One ... two ... *three*! Brilliant!
>
> Getting tired? Hop into the circle and fall down! Hop up and fall down again. Relax.

In this simple exercise the children engage in basic dance training: bodily awareness and coordination doing the hops, standing, sitting and rising; spatiality, rhythmicality with stops and force, all done together, in community. They interplay with the space-enforcing props in different colours laid out over the floor. Each dancer has their own hula hoop that they hop in and out of on the beat of the drum and the rhythm of your voice. Next time the hop theme comes up the children can make hop combinations of their own choice and explore new variations with different parameters.

The children also engage in several essential mathematical activities according to Bishop's (1988) six categories: play, explanation, localisation, design, measurement and counting.

Even though this dance play is steered by the dance leader, the children go along with the setup of the *play*. They think of the hula hoop as a *circle*, an abstraction of something familiar to them from their everyday reality. They perform and explore habitual movements in a formalised ritual with rules they can predict after the first trial and evaluate what the next step will be. Playing the game, they handle positions, the numbers one to three and the circle's form, all in a social situation with their teacher and peers.

Asked questions why it is called a *hula hoop* and what *round* means, they can reason and *explain* the drum is round and describe round forms. This *meta-talk* after the dance activity offers opportunities to ask more questions. Putting words to everything with your help, their explanations lead on to elaboration and an increased understanding of various forms around us as well as an increased ability to draw conclusions as to whether different forms are round or *circular* like circles or cornered or *angular* like squares: they engage in mathematical researching. Ask them to look for circles in the room, to describe the shape of clocks, sockets, screws, buttons on clothes. Look each other in the eyes, count all the circles and make drawings of eyes with irises and pupils afterwards.

The *localising* activity coincides with the dancing relating to the surrounding space. Also here questions such as *where*, *whereto*, *in which direction* and other positioning terms help the children *see* and *code* the room, in relation to themselves, to the hula hoop/the circle, to the floor area and the room space and to all the others around. The dance displays all these positions, *teaches* us by offering a chance to kinaesthetically experience all the movements between the different positions: in and out, down and up, forward and backwards.

When doing the dance exercise the children do not engage in *design* as mathematical activity. On the other hand, the dance experience may stimulate further activities such as manufacturing and shaping round objects and other geometrical forms, and using these as additional props when dancing and discussing geometry. The children can shape, reshape and arrange pegs, clay, plastic and stones, and describe their appearance and forms again: a circle, a sphere, a square, and all their properties.

In connection to the dance exercise the children can also engage in *measuring* if you hand out more rings, circles and ropes they can lay in circles and put them inside or outside the hula hoops. They can compare sizes, amounts and lengths, the shape remaining constant. The objects can also be ordered according to quantity, for the children to observe patterns. Hand out ropes of different lengths, let them work in pairs, one standing and holding the rope in one end, the other one holding the other end and dancing round the one standing and spinning round on the spot. This offers another kinaesthetic experience of the circle, of its *centre* and its *radius* whose length determines how big the circle is.

Finally the children dance and *count* in the play. They wait, count one, two, and on the three hop in or out, up or down, forward or backwards. Here the counting is performed in an explicit way, following an instruction. But all dancing is a matter of counting. Children who gallop or skip spontaneously dance at an uneven 6/8-rhythm—or simplified in a duple time—and the rhythm is manifest to both the dancer and the beholder, whether one can count explicitly like a musician or not. Kinaesthetic ability precedes the cognitive. The notions do not necessarily need to be explained before they can be handled; nonetheless in the dance they are emphatically handled and even filled with meaning. They are experienced kinaesthetically and affirmed with appropriate language in the discussions. Dance teaches counting, geometry, mathematics.

Relate this counting to the numbers 1, 2, 3 and on in the Roman, Arabic and Indo-Arabic number systems, stressing once again the significance of language in math. Count, sing and dance with the children in all their mother tongues.

The Circle—for Not So Young Children

Music: driving, steady rhythm, e.g. Omar Faruk Tekbilek: Imaginary Traveller.

Choose those parts from the dance and the discussion below suitable to the group at hand. Call out *circle* again.

> Everybody dances into a big circle. Join in and help keep the dance flowing. Ask someone to dance in the circle's *centre* or *middle point*. Someone goes dancing in the middle. Everyone dances like the dancer in the centre.

Who can dance a *radius*?

> Someone dances to the centre and back to their place in the circle. Or someone goes and stands with one arm stretched towards the centre and one arm towards their place in the circle. Or a few put themselves in a line between the centre and the circle curve. Ask some-

one else to dance a radius, and another one. Ask everyone to dance 'their' radius all at the same time – everyone dances to the middle and back, all the time doing the centre's moves.

Who can dance a *diameter*?

Someone dances to the middle and all the way to the opposite spot on the circle curve. Or some shape a line through the centre. Ask someone else to dance a diameter. Ask everyone to dance a diameter all at once. Everyone dances to the centre and on to the opposite position in the circle.

Can anyone dance the circle's *circumference*?

Someone takes off in a loop around the circle. Everyone else stays in their position on the circle curve and dances like the one dancing the circumference. Ask them to check where the circumference dancer starts and finishes: one round around the circle! Dance like the centre dancer again!

Go round inside the circle, point and give everyone a number: one, two, one, two, one, two. All the *ones* dance the circle's *area*!

Dance and make sure to cover all the area inside the circle. The *ones* go back to the circle, then the *twos* step in and dance the *area*. Ask the circumference dancer to dance it again while the *twos* dance the area.

Everyone dances back to their places on the circle curve and finishes dancing like the centre dancer, who gets a round of applause for an excellent job in the service of dance and geometry.

Discussion and Formalisation

Turn of the music. We are standing in our circle. We have danced the circle and a great deal of its properties. Now let's put words to it all. What is a circle, how do we define that geometrical form? *A round ring. A round curve.* What does *round* mean? *It has no edges. It just goes round, round.*

Let's sort it out.

Draw a circle on the board or ask someone to do so and fill in everything we state about the figure. Discuss what *the centre* is. A point in the middle, *equidistant* or *equally long* from all the points out on the circle's round line, the circle's *circumference*: a circle is a plane shape of all points at the same distance from a given point, the *centre* or *origin*.

Ask the centre dancer to go and stand in the middle again. Mark the origin in the circle on the board with an O. How long is it from every one of us to our dear centre? A *radius* distance. Draw the radius, mark it with an r.

Take a rope. Make a knot in the one end and ask the centre to hold it. Take your place in the circle, stretch the rope between the centre and yourself, make a knot in your end. Let the knot go round to everyone, who all adjust their positions in the circle, getting a strong spatial and corporeal perception of the circle's perfect rounding. Point out the rope's, the *radius's* straight line. What happens if the radius gets shorter? Or longer?

Take a step forward, making the rope shorter, everyone moves inward, the circle gets smaller. Draw out the rope again, everyone backs out again, the circle gets bigger.

Discuss the *diameter*. A straight line through the centre, and its length. Draw a diameter in the circle on the board, call it *d*. How many times the length of the radius is the diameter? Show with the rope if necessary. Refer to the notion *double*, which can be applied to length and weight as well as numbers. Write the formula on the board: $d = 2r$.

Discuss the *circumference* someone danced. One circuit around the circle: the very *curved line* the circle consists of. Draw the circumference on the board in a different colour and mark it with a C. Take a rope the length of a diameter rope and measure how many times it fits one round around the circle, its circumference: three times and some more. Tell about the ancient Egyptians who studied this *ratio* in every single circle and established the relation between the *diameter* and the *circumference*: the *ratio* was found to be slightly *3.1416*, a constant that can never be determined exactly. Eventually it was designated by the symbol π – *pi* – from the Greek word περιφέρεια, *periféreia*, for circumference. Write the formula on the board: $C = \pi d$ or $C = 2\pi r$. Thanks to the ancient Egyptians and the Greek we can calculate a circle's circumference instead of measuring it with a rope or a rod. Isn't it intriguing to think how π had been hiding there in the abstract but definitely real world of math, waiting to be discovered by inquiring minds?

Discuss the *area*, the surface enclosed by the circle's curved line, its very *circular disc*. Let someone fill in the circle's area on the board with a colour. Denote it as *A*.

How is the area of a circle calculated? $A = \pi r^2$. That too we know thanks to a clever old Greek, Archimedes. (In the next chapter we will imagine taking a bath like he did, but that's another matter, a weighty and physical one.) The circle's area is the sum of the areas of a *countless* number of *infinitely* small triangles, like super thin pieces of cake one next to the other: each triangle has one corner in the circle's centre and two equally long sides as long as the radius *r*. The third side *a* is the shortest line you can imagine between two points on the circle curve, almost as if they coincided. That way these infinitely small triangles are right angled, so their small areas are $\frac{1}{2}\, r \times a$ (see below: Triangle). The sum of these exceedingly small triangles is $\frac{1}{2}\, r \times$ the sum of all the exceedingly small sides *a*, that is one loop around the circle, that is … the circumference. Thus the circle's area is: $A = \frac{1}{2}\, r \times C = 1/2\ r \times (2\pi r) r = \pi r^2$. This will become as clear as day later on, when we dance all polygons and line up all fractions and blow our minds on the enigma of the multiplication by 0 of a number approaching innumerable …

Ask them to point at other circles in the room: the clock, sockets, air gratings. Don't miss buttons and dots on sweaters, eyes, irises, pupils. Extend the dance session by doing drawings and colouring all the geometrical notions and their relationships, while the moving images of the dancer in the *centre*, the one dancing the *circumference* and all dancing the *area* are still fresh in mind.

The Ellipse

Everyone is standing in a circle again. Reaffirm the circle as a plane geometric figure with a round curved line, the *circumference*, where every point on the line is at equal distance, the *radius*, from a fixed point, *the centre*.

What then is an *ellipse*? It is a plane closed curve too. Take a long rope. Ask two pupils to go inside the circle and position themselves on a diameter a bit apart from each other on either side of the centre. Give them each an end of the rope. Ask a third pupil to step in, pick up the rope and keep it stretched between themselves and the two in the middle, shaping a triangle. Ask the three to hold the rope a bit above their heads, the third one holding the rope tensed with an index finger or a pin. A fourth pupil steps in next to the third one and takes over the rope with a finger or the pin, holding the rope tensed between themselves and the two in the middle, shaping a new triangle. Another one steps in, takes over the rope, forms a new triangle, passes on the rope to the next one stepping in, and so on. Bit by bit the circle reshapes into an ellipse. Two of the pupils stepping in end up in line with the two in the middle, the rope also forming a line; opposite each other, these two stand where the ellipse is at the flattest.

A more dancing way to establish the ellipse is to ask one pupil to hold up the rope with a finger or a pin, keep it tensed between themselves and the two in the middle, and dance the entire circumference of the ellipse.

So what is an *ellipse*? Like the circle it is a plane curved line. Here the constant is the *sum* of the distance between each point on the curved line and *two* fixed points, *focal points* or *foci*. The two pupils in the middle represent the foci. The rope's length, the sum of the two sides of the triangle shaped by the rope between them and any pupil on the ellipse's curved line is the constant.

Ask the foci to increase the distance between themselves, and check how the ellipse narrows. Ask them to bit by bit move closer to each other and see how the ellipse gets rounder and rounder. When does the ellipse become a circle …? When the focal points coincide! What is the constant then? Twice the radius, the circle's diameter! So, a circle is a special variation of an ellipse, where each point on the curved line is at a constant total distance, a diameter's length, to two imagined points who coincide.

Are there other geometrical figures which are special variations in different categories …? Yes, obviously. Now we're into variation theory, so come and swing along.

The Rectangle and the Square

Turn on the music, a driving rhythm. Everyone dances freely. Call *rectangle*! Everyone dances into a rectangle. Let it dissolve, ask for different rectangles: long and drawn-out, small, two, three. Then ask them to shape one big rectangle again.

Draw or ask someone to draw the rectangle on the board.

How many *angles* are there? The four angles hold up their hands. How many *sides*? Four. Draw a *parallelogram* on the board. Is this a rectangle? Not quite. It does have four angles and four sides, so it is a *quadrangle*. How are the angles in the rectangle? *Right* angles, 90°. That makes the opposite sides of the rectangle, as in the parallelogram, *parallel* and equally long. What are the sides called? *Length* and *width*. Mark two *l*s and two *w*s in the figure on the board. Ask the group to make the rectangle smaller, then as large as possible in the room. How many such large rectangles are there in the room? Two, the floor and the ceiling—that is, if the room is ... *rectangular*.

Ask someone to dance a *diagonal*. Point out that it goes between two opposite angles. How many diagonals are there? Two. Ask two pupils to dance both diagonals and see where they cross each other—a *middle point*, different from the circle's *centre*. Mark it on the board.

Does the rectangle have a *circumference*? The figure's length around the outside is called the *perimeter*. Does it have an *area*? Ask someone to dance the perimeter. Ask the previous number ones to dance the area, like in the circle, then the twos. How do you calculate the rectangle's perimeter and area? Formalise on the board: $P = 2\,(l + w)$, $A = l \times w$.

How do we transfer a rectangle into a *square*? The dancers readjust the right-angled quadrangle into a square. Draw it all on the board. How many angles are there? Four. How many sides? Four. Write an *S* by each *side* on the board. What is the same and what is different compared with the rectangle? Compared with the parallelogram? A square is a special rectangle which in turn is a special parallelogram. How many diagonals are there? Two. Someone dances the square's perimeter, the *ones* dance the area, then the *twos*. How do you calculate the perimeter and the area? Refer to the rectangle. Write the formulas on the board: $P = 4S$, $A = S^2$.

A Square and a Cube with Half as Long Sides

Repeat the terms *angle, side, right angles, equally long sides*. Go through the notions danced: *diagonal, midpoint, perimeter, area*. Repeat the calculation of perimeter and the area, show it on the board: $P = 4S$, $A = S^2$.

Ask two pupils to stand each in the middle of one of two sides of the square shaping a corner. Ask some of the others to position themselves in a small square inside the big square, with one angle in common and the opposite angle coinciding with the big square's midpoint. The other two sides of the small square go from the big square's midpoint to the middle of its sides, where the first two pupils stand.

Now we have a big square and a small one inside whose sides are half as long. Draw the figures on the board. Name the big square's sides *S* and the small square's side's *s*. Write the ratio $S = 2s$ and $s = \frac{1}{2} S$.

Ask someone to dance the perimeter *P* in the big square again. Then ask someone to dance the perimeter *p* of the small square. How much shorter is the small perimeter compared to the big one? Worded differently, how many times does the small perimeter p fit into the big perimeter P? *Two times!* Reason together, make use of the figures on the board, and write the formulas $P = 4S$ and $p = 4s$. $P = 4S = 4\,(2s) = 2\,(4s) = 2p$.

Ask some to dance the area in the big square and then the area in the small one. How much smaller is the small area a compared to the big area A? In other words, how many times does the small area fit into the big area? *Four times!* Write the formulas $A = S^2$ and $a = s^2$. $A = S^2 = (2\,s)^2 = 4\,s^2 = 4a$.

All of this is revealed in the dancing, a kinaesthetic, spatial and visual experience of how the small square's perimeter obviously is half as long as the big square's, and its area a fourth of the big square's area.

Imagine now, starting from the big and the small square on the floor, a big cube with sides S and a small cube with sides s. Use sticks the size of S and s for additional spatial and visual support. How does the volume V in the big cube relate to the volume v in the small cube? Put differently, how many times does the small cube fit into the big one? Eight times, obviously. Write the formulas on the board: $V = S^3 = (2s)^3 = 8\,s^3 = 8v$.

Remind everyone of the notion *dimension*. When a geometrical figure is made double as big, all one-dimensional measures are doubled: all *linear* measures or measures *of length* become twice as long. All two-dimensional measures, *square* measures or measures *of area*, are doubled once again. Two times two times makes four times. And all three-dimensional measures, *volumetric* measures or *cubic* measures, are doubled once again. Two times two times two times makes eight times … Inversely, when a geometrical figure is halved all linear measures are halved, all square measures are halved once more to a fourth and all cubic measures are halved once again to an eighth. Write the ratios on the board:

$s = \frac{1}{2}\,S;\ p = 4s = 4 \times \frac{1}{2}\,S = 2S = P/2;$

$a = s^2 = (S/2)^2 = S^2/4 = A/4,$

$v = s^3 = (S/2)^3 = S^3/8 = V/8.$

The Triangle

Put on the music again. Everyone dances freely all over the floor. Call out *rectangle* again. They shape a rectangle. Ask some to form a diagonal, dividing the rectangle into two figures. What is that figure called? A *triangle*. Ask the two halves to move apart a bit, form two identical triangles. How many angles and how many sides does a triangle have? Three. Ask some to dance the *perimeter* of a triangle and its *area*. How do you calculate the area of a triangle? Relate this to the rectangle we started from. The areas of the two triangles are half the area of the rectangle. Discuss the triangle's *base* and *height* in relation to the rectangle's length and width. Draw everything on the board, write the formula for the triangle's area: $A = \frac{b x h}{2}$.

Discuss *congruent* triangles, such as the two identical halves of the rectangle: you can lay them exactly on each other. Note that the two triangles are *right-angled* as the rectangle is *right-angled*. Mention other aspects, like when a line is drawn inside the triangle *parallel* to the base, then two *uniform* triangles are shaped, the original and a smaller one with the parallel line as its basis, with one common angle and all angles equal.

Remind them, as we did with the square, that geometrical figures can be doubled and multiplied, transferring sides, perimeters, areas and volumes accordingly to the multiplication.

The next dance teaches *isosceles* and *equilateral* triangles.

Eternal Isosceles Triangles

Music: rhythmical, gently driving, e.g. Souad Massi: Ghir Enta.

Everyone is spread all over the floor. Each dancer picks in secret two partner dancers. Throughout the dance everyone will try to keep the same distance to their two secret partners, while each of these two also try to keep the same distance to their own two secret partners, who in turn keep an eye on their two secret partners, who in turn …

Put on the music. Encourage lively dancing, big and smooth arm and leg movements, turns and leaps while adjusting positions according to how your two partners move. Steadily shifting movement patterns arise in the room. Many secret eternal triangles occur in the room …

If it gets tricky when your two secret partners come very close to you, or you find yourself on a line between them, then cross through the line, spin round, move away a bit and continue dancing, keeping your distance to both of them.

The dance may go on for a while. Sometimes the flow eases down as everybody lands in their right positions. Then maybe someone makes a minor displacement disturbing the symmetry for someone else, which in turn impacts on others' positions, and the flow of movement gains momentum again. Sooner or later everybody settles in the right position with regard to each other. Sometimes this state is achieved after just a few minutes. A magical state.

Turn the music off. Time to reveal those involved in all the eternal triangles. Ask everyone to unveil their secret partners. Use ropes or ribbons in different colours to help show this. What *figures* were shaped in geometrical terms? *Triangles!* In which way are all these triangles special? They are *isosceles*: two sides are *equally* long. Each dancer is the vertex of an isosceles triangle where the two equal distances compose two equally long *sides*. The line between the partners constitutes the *base* of the isosceles triangle.

Check if there happens to be a triangle among all these isosceles with *three* equally long sides. How is such a triangle named? *Equilateral!* Check also the occurrence of *right*, *acute* and *obtuse angles*.

Polygons

Summarise the various geometrical figures we have met so far, all going out from *point* and *line* to *plane* and *space*: triangle, square, cube, rectangle, circle, ellipse. Expand on to *polygons*—the *pentagon* with five sides and angles, the *hexagon*, or *heptagon*, the *octagon* and so on, all formed by a Greek number and the suffix *gon* for *angle*. The prefix *poly* means *many*. Discuss and clarify how many sides and angles comprises each figure. Find different forms and figures in the room, draw attention to houses, façades, show pictures of buildings, sculptures, paintings, films.

Repetition with Dance Stops

Put the music on again, everybody dances. Stop the music, call out a geometrical figure or a notion:

> Dance the longest diagonal in the room! Form three circles! Indicate their centres, their circumferences. Shape a square! A pentagon! Two *dodecagons* (if there are 24 dancers!) Line up along the other diagonal in the room!

Parachute Dances

All Geometry in Five Minutes

Music: steady rhythm, disco, pop, rock, techno, world.

The parachute with its geometrical form and group gathering effect is a useful prop in dancing math content, such as geometrical figures, fractions and infinitely big and small numbers.

In the following we dance through the entire geometry in five minutes: all polygons all the way to … the circle. Then we take the opportunity to take on all fractions. Then we tackle the enigma of the multiplication with zero of a number striving towards *innumerable*.

Concluding the math chapter we look at the clock and take some time to look back at Roman numbers, sensing the pulse of a number system with basis 60.

Everybody is standing in a ring. Inside lies a colourful parachute, crumpled up in a little bundle. Remind the dancers about what geometry is. Math you can draw, math in space. Points, lines, planes, and all figures and bodies you can shape with these elements: circles, rectangles, polygons, cones, cylinders, spheres. In this dance we stick to two-dimensional, plane figures, apart from the very beginning.

Put the music on. Everybody rocks and dances on their spots around the ring.

What is the tiniest smallest thing one can draw? A dot! In math language: a *point*. Ask someone to pick up the parachute bundle, hold it in their arms and dance inside the ring. Everybody standing in the ring dances like the dancing *point*.

Ask another *point* to step inside the ring and dance with the first one. The two points unfold the parachute into a strand between themselves while keeping up the dancing. Everybody dances along. What is it you can draw between two points? A *line*! A connection between two points. Step in, dance a long winding line between the two points and ask if that is a line …? Sure it is, a *curved* line, an awfully long one. What is the *shortest* line between two points? A *straight* line. Ask the two dancing points to hold the parachute fabric tightly stretched between them, like a straight line. Keep in mind a line consists of lots of points next to each other on a row, with no width nor depth: a one-dimensional figure.

Ask a third point to join in. The three points unfold the parachute and stretch it out between themselves. What geometrical figure do we get? *A triangle!* How many sides are there in a triangle? Three. How many angles? Three, our three dancing points. Ask the three dancers to shape an *equilateral* triangle.

As more points team up, we make sure all dancing points are at equal distance from each other. That way we get *regular* figures, with equally long sides and equally big angles. Keep in mind that every plane figure we shape is made up out of straight lines on a plane. If we would move on from *plane geometry* to *space geometry* and three-dimensional figures, we would see that they in turn are shaped by planes. For now we stick to plane geometry.

A fourth point dances into the ring. The four points stretch the parachute between them, keeping the same distance between each other. What geometrical figure do they form? *A square!* If someone says a *parallelogram*, reaffirm it could be the case. Ask the four, keeping an equal distance between each other, to also make sure the angles are equal: *right* angles or 90°. So we get a square.

A fifth point dances into the ring. The five unfold the parachute between them and spread evenly around it. Which geometrical figure is this? A *pentagon*. How many angles does a pentagon have? *Five*. How many sides? *Five!* A sixth point dances in. What figure do we get? A *hexagon*. How many angles, how many sides? *Six!* Keep the rocking and the dancing going so the build-up evolves in a flow easy to follow and predict. Another point joins in, and another one, and another. A *heptagon* comes into being, an *octagon*, and so on. Keep asking how many angles each figure has and how many sides: seven, eight and so on. Point out how the sides get shorter and shorter as more dancers come in, spreading equally around the parachute.

Go on to the *nonagon*, the *decagon*, the *hendecagon*, the *dodecagon*. The parachute usually has twelve *segments* or 'pieces of cake'—we'll come back to cake pieces later. What do we have so far? A twelve-sided *dodecagon*. How many angles? *Twelve!* How many sides? *Twelve!*

Another *twelve* dancing points join in, taking place in between the twelve holding the parachute. What do they shape together? A twenty-four-sided polygon or *icositetragon*. How many angles and sides? *Twenty-four*. All the remaining dancers move in, including yourself. Everybody positions themselves equally divided around the parachute. Turn the music off.

Now we are many dancers shaping a figure all together. What do we call a geometrical figure with many angles? A *polygon*. Which polygon do we shape with our group, our class? A *twenty-six-sided* polygon or a 26-gon, a *thirty-sided polygon* or *triacontagon* or a 30-gon …

If we were fifty, how would that polygon be named? A *50-gon* or *pentacontagon*. How many sides, angles? *Fifty*. If we were a *hundred*? A *100-gon* or *hectogon* or *centagon*, with a *hundred* angles and a *hundred* sides. Imagine how short the sides would get if we were a hundred holding the parachute. What if we increased to five hundred, to a thousand? A *500-gon*, a *1000-gon* or *chiliagon*, with *five hundred* respectively a *thousand* angles and sides. If we were a million? A million-gon or *megagon*. A *billion*? A billion-gon or *gigagon*. Ask everyone to crowd closer and closer to each other. The sides of the polygon become shorter and shorter. Angles stand thronging one onto another. In the end there is hardly any space left between the angles.

Now imagine that all the children from all the schools in the world, a couple of billion, hold the parachute. Think of a thousand billions, and billions of billions from all planets where there may be schools and children or ETs.

What did we call the number that keeps increasing and increasing, getting so big you can no longer count it no matter how long you kept counting? *Countless*, *innumerable*, *infinite*, symbolised by ∞. A number which only exists in thought. It is no real number, but a undefined number. A number that is not counted among the *real* or *rational* numbers: it is an *irrational* number. Just like the geometrical notion *endless*: a distance endlessly far away, a line drawn out into space towards galaxies far, far away and still further onto *endlessness* or *infinity*—it never ends.

If more and more and more people keep joining us around the parachute, then ultimately there is no more room between our points, the sides get more and more numerous and shorter and shorter, tending to zero. Which figure do we get at long last? *A circle!* We can think of a circle as a polygon with *innumerably* or *infinitely* many angles.

Note that this is the case in our dance as we hold the parachute and jam more and more dancers/angles around it; then the perimeter of the inscribed polygon tends to merge into a circular perimeter. If we would instead imagine gathering on a prairie or in a desert and bring in more and more people and extraterrestrials, moving further and further away from the centre while the number of sides of the polygon we shape between us increases, we would be shaping an *apeirogon*: a *boundless* polygon with infinite angles and sides.

All Fractions

We are standing evenly spread around the parachute, holding it tensed between us and pondering on *polygons* and *the circle*. Now think of the parachute as a cake. Imagine it's a girl's birthday and she throws a cake party. But a snowstorm strikes so hard no one can go out. No one comes to the party, the girl is all alone. How much cake does she get to eat—for comfort? A *whole* one. The cake is divided by 1, it remains whole or *undivided*.

Write the *division 1/1* on the board. The girl eats one whole cake by herself. Write the multiplication 1 × 1/1 = 1 on the board. Continue to write all fractions and multiplications on the board.

The neighbour boy finds a shovel and shovels an alley to the birthday girl's home. They have a party of two and share the cake. How much of the cake do they get each? *Half* a cake: ½. How much cake do they eat together? A whole cake: 2 × ½ = 1.

A friend a bit farther away owns a helicopter, so they get to be three at the party. How much of the cake does each one get? A *third*: 1/3. How much cake do the three eat together? A whole: 3 × 1/3 = 1. Another friend manages to come on skis, now there are four at the party. How much cake does each one get? A fourth part or a *quarter*, ¼. How much cake do the four eat together? A whole cake: 4 × ¼ = 1. Is a quarter of a cake smaller or bigger than a third or a half? *Smaller!* Of course.

Snowploughs start clearing the snow, five children make it to the party. Each one gets a *fifth* of the cake, 1/5, together the five have a whole cake: 5 × 1/5 = 1. If six make it each one gets a *sixth* part of the cake, 1/6, and together they eat a whole cake: 6 × 1/6 = 1. The cake pieces gets smaller as the cake is divided into more pieces. Note the fractions and multiplications on the board. Two things we can observe. First, a division is a reversed multiplication: *division* and *multiplication* are each other's *inverse*. Second, the larger the number you divide by—the *denominator*—in fractions with the same *numerator*, in this case 1, the lower the fraction's value. The more children who make it to the party the smaller each one's share of the cake is. But shared joy is multifold joy. Back to the cake. Now all classmates manage to get there, 24 cake lovers. Everyone gets a *twenty-fourth* of the cake: 1/24. Together they eat 24 × $^{1}/_{24}$ = 1, one whole cake.

The sun shines, the snow melts and rumour spreads about the yummy cake with its scent. A hundred children come who get a *hundredth* of the cake each, a tiny little tasty *1 percent*, 1/100. Nonetheless the hundred children eat 100 × 1/100 = 1, a whole cake. If a thousand children show up the slice each one gets fits on a child's pinkie nail, a *thousandth of a* piece of cake, *1 per mill*, 1/1000. Still together they eat a whole cake, 1000 × 1/1000. Continue the thought experiment with a million children, a billion. The pieces of a *millionth* and a *billionth* of the cake go on the tip of the hair on a pinkie …

The Enigma of the Multiplication by Zero of a Number Approaching Innumerable

Conclude the thought experiment imagining *innumerably* many kids coming to the party, crowding around the yummiest cake ever baked. It gets to be like the parachute circle in the earlier exercise. As in a circle where one point stands next to the other with ultimately no space left in between, the kids stand around the cake. The cake pieces ultimately *vanish* ... they become null and nothing. Dividing the cake among countlessly many birthday party guests would be like writing the fraction $1/\infty = 0$. However in our thought experiment they would all together still eat a whole cake, which would be like writing $0 \times \infty = 1$.

This would be a very naughty thing in mathematics to say or write. Only *real*, *rational* numbers can be multiplied by each other, including by zero. The same goes for division: a number, 1 or any other, can only be divided by other real or rational numbers.

All the same this train of thought is creating a logic, and as such it must be given prominence in all playing and learning contexts. Imaginary play is part of the whole meaning of dancing, of playing, experiencing and learning math in this way. Consequentially this play in thought and movement belongs to Bishop's mathematical activity *play*. When we think of the parachute as a circle and then as a round cake, we think hypothetically: think if no children show up, if only one comes, one more and one more, if a hundred come, a billion, *innumerably* many.

For sure this is abstract thinking including *modelling*: abstracting aspects of reality, like a birthday party, formalising and ritualising according to a clear mathematical procedure where we bit by bit predict, guess, evaluate, assume and *explain* what happens if more and more guests show up. We explore and stimulate the understanding of numbers, real and unreal ones; we acquaint ourselves with numerous geometrical forms, with fractions and wholes, division and multiplication as each other's inverse, during an ongoing fictitious story with definite, watertight mathematical arguments.

The issue of only real numbers being handled in arithmetic operations is formally solved in mathematics by introducing the notion of *limit values*. In our thought experiment we divide 1 by a number a, and let the number *approach* or *tend to* innumerable. The value of the fraction gets smaller and smaller, *approaching* the limit value 0. In formal math language it is written as $\lim_{a \to \infty} (1/a) = 0$ and read as: *the limit for the division of 1 by a as a approaches infinity equals 0.*

As to the multiplication of 0 by a number *approaching* infinity, mathematicians are still racking their brains whether a similar inverted formula may be written and if one may say that the limit value of the product would be 1. That is an enigma for artists, philosophers, dancers and children dreaming of being mathematicians to keep pondering about.

The Clock

A parachute with twelve segments also comes in handy in other math dancing and playing. The parachute lies unfolded on the floor. Everybody is sitting in a circle around it, leaving space so that some can move around it in a while.

> Distribute plastic coated numbers 1 to 12, which the children lay out round the parachute like on a dial. Use Roman numbers from I to XII like on old clocks or church towers and reconnect to the Roman number system and way of counting.

What more do you see on a clock face beside numbers—those things actually *indicating* or *pointing* at what time it is? *Indicating hands* or *pointers!* How many are there? *Two!* What are they called? *The big hand* and the *small hand. Three!* Yes, often there is also a long and thin pointer. Unlike the big and small hand it moves so fast you can observe it with the naked eye. It's the *second pointer*. We'll come back to this.

Let's say it's three o'clock. Who wants to do the small pointer? Where does it point at three *sharp*? Someone lies down, feet to the middle and head to 3 or III. Who would like to do the big pointer? Where does it point when the clock strikes three? Someone else lies down stretched out full length on the parachute, feet to the middle, arms or head to 12 or XII.

Now we let the clock tick from three to four o'clock. How much time does it take? *One hour!* How does the big hand move during the hour between three and four? *One round!* Our clock is fast, we don't have all the time in the world and neither do we need that much time to learn the clock. It takes one hour for the big hand to go one round from XII to XII again. Take a minute and think about how many minutes it takes—how many minutes are there in an hour? *Sixty minutes!* The big hand indicates them, minute by minute, and is called *the minute hand*.

How does the small hand move in the meantime, between three and four o'clock? Everyone points at four. Where did the small hand point at three o'clock? At III! And now, at four o'clock? At IV! While the big hand moves from XII to XII again the small hand goes from III to IV. If the big indicator indicates the minutes, then what does the small hand indicate? First it pointed at the hour three, then at the hour four. It indicates the hours. The small hand is *the hour hand*. Time goes on. What time will it be one hour from now if now is four o'clock? *Five!* We ask the big and the small hand to move from four to five o'clock. The big hand goes one lap again from XII to XII, the small hand shuffles from IV to V. They *time* each other to start and finish at the same time.

Discuss different things that take an hour: a lunch break, an episode in a tv-series, the bus ride to the beach.

Our clock is ticking. It's five o'clock, what time will it be in half an hour? Five and a half …? In many languages, yes. *Half past five* or *five thirty* in English. How do the hands move from five o'clock to half past five? The big and the small hand on the parachute move. Where does the minute hand point to half past five? Down to VI, halfway to XII. The hour hand? Half-way between V and VI.

Time never stands still, so we let the clock go another quarter of an hour. What time will it be then? *A quarter to six.* How do the hands move? The big hand goes another quarter of a round, to IX: one quarter of a round left towards XII for six o'clock. The small hand? Another quarter of the space between V and VI.

Discuss *ten to six, five to six, a minute to six, five past six, six twenty* and *six twenty-five*. Summarize the terms hours and minutes. So how many minutes are there in an hour? *Sixty!* How many minutes does it take for the big hand to go one round from XII to XII? *Sixty!* How many minutes are there in half an hour? *Thirty!* A quarter? *Fifteen* minutes! Three quarters? *Forty-five* minutes! How many minutes does it take for the big hand to go from XII to III? *Fifteen!* From XII to I? *Five*—first one minute after the stroke of the clock, then two, three, four, *five*! And on from I to II? Five more minutes: six, seven, eight, nine, *ten*. Every bit between the numbers on the dial is five minutes. Discuss things that take five minutes: boiling an egg, dancing to a little longer tune …

Remind them of the 5 times table: $1 \times 5 = 5$; $2 \times 5 = 10$; $3 \times 5 = 15$; $4 \times 5 = 20$; $5 \times 5 = 25$; $6 \times 5 = 30$; $7 \times 5 = 35$; $8 \times 5 = 40$; $9 \times 5 = 45$; $10 \times 5 = 50$; $11 \times 5 = 55$ and $12 \times 5 = 60$.

So what about that third pointer on many clocks? It's the *second hand*. How long is a second? Count *one (thousand), two (thousand), three (thousand)* … How many seconds are there in a minute? *Sixty*. As many as minutes in an hour. The clock is a base-60 number system.

> Let's walk a whole minute like a second pointer. You may have seen one of these clocks on a train platform or at an airport, where a red dot lights up on top of the dial right of 12, and then another one lights up and another, one new red light for every second all the way around to the sixtieth when they turn off and it starts again initiating the next minute to tick away second by second.
>
> Who wants to go round our clock like this from XII to XII on one minute? Everyone except the three hands rises to stamp the pace and count while some move around the clock. Do you remember how we stamped the five times table? How was the rhythm? One two three four *five*, one two three four *five* … stamping hard on every fifth number. What did we start on? Zero! How long did we count? To 55! Do you manage to count to 60?
>
> One, two or three dancers go and stand by XII, ready for one lap around the clock dial. Everybody stamps the beat, one stamp a second … ready, steady … *Zero* 1 2 3 4 *five* 6 7 8 9 *ten* 11 12 13 14 *fifteen* 16 … 48 49 *fifty* 51 52 53 54 *fifty-five* 56 57 58 59 *sixty*!

Do the minutes' lap again with twelve dancers positioned around the parachute. Children offer many suggestions on how to do the pointers. A tall one and a less tall one do the minute and the hour hand. Four or five move round shoulder to shoulder doing the big hand while two or three do the small hand. One stands in the middle and points with an outstretched arm as the big hand, holding the other arm's elbow close to the body and pointing as the small hand. Or someone does the second hand stepping round the parachute like the red dot moving on the dial on the train platform.

Conclude by affirming that what we've just done is *measuring* time and that we did so by using a number system with base 60. Relate the clock theme to language idioms or metaphors: *to tell the* time; *to know the clock of the day*; *to arrive at the eleventh hour*; *to be back in a second / in a sec / in half a tick*; *time and tide wait for no one*; *to work round the clock*; *lost time is never found again*; *to turn back the clock*; *better three hours too soon than one minute too late*; to hear something so stupid *the clocks stop*; *to be so bored the clocks go backwards*; *their time is numbered*; *even a stopped clock is right twice a day.*

And, from The Little Prince by Antoine de Saint-Exupéry: *It is the time you have wasted for your rose that makes your rose so important.*

Relaxation

Odd or Even

Music: calm, sea rush, winds, e.g. Radio Tarifa: Soledad; prop: big fan; see description of the basic dance movements in 'Rocking and Rolling Waves' in Chap. 2.

Everybody is lying on their back spread over the floor, eyes closed, feeling their heart pulse. We imagine we are lying on a sandy beach relaxing. It's summer, we feel the fan of a fresh breeze sweeping in from the water over our bodies. Go round with a fan.

> Rock your bodies gently from side to side. The breeze increases, it makes you roll over on your stomach, roll back on your back again.
>
> Listen carefully. Roll over when I say an even number. Lie just still when I say an odd number.

Start with low numbers: 2, 4, 8, 5, 3, 0, 6, 10, 11, 14 … Say higher and higher numbers: 18, 21, 53, 98, 100, 505, 3084, 102 305, 450 321 876, nine billion eight hundred and fifty-seven million nine hundred and eighty-five thousand six hundred and fifty-two …

What determines whether a number is odd or even? The last number! So just lie still and wait until I've rattled off the number and listen to the last digit if it's odd or even.

The Three Times, Five Times, Ten Times, All Times Tables

Do the same exercise with the times tables. Everyone rolls over on the numbers in the three times table, the five times, ten times table and so on.

Chapter 6
Dancing Natural Sciences

P. Moerman, *A K-12 Dance Method to Teach Creative Dance, English, Math and Science*, https://doi.org/10.1007/978-3-031-93156-7_6

Kinaesthetic and Scientific Literacy in Interplay

Dancing Natural Sciences, similar to *Dancing Math*, is a dance and natural science teaching and learning method starting from dance and the (kin)aesthetic experience of science subject content (Moerman, 2016b). The method epitomises aesthetic teaching, recognising the primary work of the senses in being in, perceiving and learning about the world, prior to—and enabling—thinking, interpreting and understanding the world.

For all intents and purposes, teaching is about turning the young individual to the world and to arise a desire to live in and with the world (Biesta, 2017, 2021). To teach is also to invite the student to attend to fields of knowledge and proficiencies, and to arise a longing to be initiated into realms of literacy: how a dancer 'thinks', how a linguist 'thinks', or a poet, how a mathematician 'thinks', a physicist, a biologist, a chemist, an astronomic.

In preschool and elementary school, subjects such as dance, language, math and science are taught in their essence and often holistically, relating to everyday life experience. Fundamental questions scientists busy themselves with are: How are things in the natural world? How does everything hang together, in space with the sun and the other stars and galaxies, with the weather and the climate, with seeds growing and becoming plants and trees, with tame and wild animals we share the earth with, with new viruses and illnesses spreading?

Even children ask these basic questions, looking up to the stars twinkling in the night sky, watching the swallows swish high in the summer sky, squatting down to glance at an earthworm slowly labouring its way into the earth, running to check how long the sunflower stalk has shot up from the seed they planted the other week. These questions own a factual dimension and one of never ceasing wonder, as well as an existential aspect related to trying to live in and with the world we were born to and interact with while borrowing and caring for it.

In this chapter dance offers a channel for these efforts to dialogue aesthetically with the natural world. Dance as kinaesthetic experience opens for an initiation into the world of natural sciences and a perception of how we are connected to it. As in the previous chapters the activities are anchored in the dance teaching method described in Chap. 2. Some dances reoccur, other dances are designed to deepen into single or sets of cohering subject content on the themes of water, air and space.

The starting point is dance. In subsequent *meta-talks* the subject science content dealt with in the dances is discussed in subject-specific language, tied to manifestations of scientific laws in everyday life and the participants' previous conceptualisations. Conclusions are drawn and the content is formalised in definitions, physical axioms and chemical formulas.

Dancing, the participants create a space of teaching and learning: a space of interaction and communication between teacher, pupils and the subject matter—*the didactic triangle*. Dance visualises aspects of subject knowledge and offers keys to scientific reflection and concept formation—dance *teaches* physics, chemistry and astronomy, and stimulates kinaesthetic, scientific and language development in beneficial interplay.

Not least, dance offers space to experience relational and existential aspects of humankind's efforts to find answers to eternal scientific questions.

Water

Warm-Up

Music: free flow, e.g. Omar Faruk Tekbilek: A Call to Prayer.

Initiate the water theme with a calm, gathering warm-up.

> Everyone walks freely over the floor, crisscross, all over. Straight backs, light steps, silent, sense the flow of movements. When meeting someone, make eye contact and turn away in a smooth bent.

Gently interrupt the flow after a while:

> Stop, stretch arms up, rise on tiptoe, stretch arms out to the sides, spin round in one direction, in the other direction.
>
> Again, walk around, at a swifter pace now, a springy but relaxed walk. Turn away when meeting someone else, an energetic but soft bent. Stop, stretch arms up, rise all the way up on tiptoe, spin, arms to the sides, in one direction, in the other direction.
>
> Same again, now on a run on tiptoe, dodge quickly when meeting another dancer. Lower the body's gravity centre slightly bending your knees, scamper about straddle-legged, stretch out arms and hands like tentacles, spread your fingers, like those insects on the water surface on the pond or in the creek—what's their name? *Water skippers!* Continue, even swifter, cover the whole floor, dodge, skip away from each other. Stop, arms up, rise on tiptoe, spin round, arms out in one direction, in the other direction.

Water Swirls

Music: water-flows, currents, rapids, waves.

Establish the water theme. Return to the smooth crisscrossing flow of movement in the warm-up. Look for a continuous flow in the steps over the floor, making the movements still smoother.

> When meeting another dancer, accentuate a smooth twist on a pivot with your arms and upper body, a gentle eddy, like two water whirls waving hi to each other. Make eye contact and, without stalling the movement, gently bend through your knees, lift your arms slowly, twist your upper bodies and arms and turn mirror-inverted away from each other with a little accent, turn your backs to each other, lower your arms, let go of the contact and move on to a new meeting, a new whirlpool greeting.
>
> Small whirls meet and exchange greetings. Give the movements a water quality, become one with the water, pleasant and light, a mild resistance in an ongoing, billowing stream.

Stop the flow of movement now and then, standing still and just billowing along in an imaginary stream on the water, nice and gently, body temperature water in movement. Find a common wave movement.

Billow on the bottom of the lake, let the stream take a hold of everything and everyone, find a joint stream movement. Find a weightless state, let the water, the billows and the currents bear your bodies. Slow down when meeting someone else, let your bodies heave and sink, sustaining the movement a second, sink away, wash on, like algae and seaweed in the current, like the water itself.

Become one with the water, gushing, streaming, whirling, speeding and breaking, move your arms in waves, let your torsos billow, try smooth twisting movements back and forth, a sudden wave heaving and rolling over, dance alternately low down and striving upward, then gently falling—improvise water movement!

Think a turbulent storm, the water gets restless, swelling and billowing in different directions, streams breaking against each other, hurling up, falling down. Then the storm abates, the seas settle. Find a common, calming wave.

Meta-talk on Water

Talk about dancing water, about the quality and feeling of movement. How does water move? How do whirls arise? How come one can be drawn down in a whirling current? Or pulled out to sea by the undercurrents?

How do striders and other arachnids manage to run over the water's surface? How many legs do they have? What do their legs look like? What is surface tension? Why are they called striders or skippers? What are they called in other languages? *Writers* in Dutch, *weavers* or *shoemakers* in Spanish, *water skaters* or *water spiders* in French, in German *water runners*, *water tailors*, or just *tailors* like in Swedish, *water-skiers* in Arabic, *water measurers* in Finnish and Russian…

Dancing on Ice Floes

Music: quadruple beat, driving, e.g. Omar Faruk Tekbilek: Imaginary Traveller.

Everyone is standing by one of the short walls, facing the room. We imagine a big river in front of us, not too rapid-flowing. School is on the other side of the river. Each morning we walk a long way along the riverbank to a bridge which we cross, and then we trot an equally long way along the opposite bank to get to school. When winter comes and the ice starts covering the water we can take a short cut over the floes. Then we must jump with vigour from one ice floe to another as they float downstream, we must keep our balance and watch out so we don't slip.

Dance across the river in groups of five or eight at a time. Our first attempts are in the beginning of winter, as the ice is just thick enough to bear us and we are not yet used to saving time this way.

Jump, totter, gesticulate violently, struggle for balance, shift between dancing on tiptoe and on the whole foot. Let it all take time doing a balancing dance across the big river.

Experiment jumping forward, backwards and to the sides, on to the next ice floe floating by. Jump back over the river on the way home.

Imagine late winter, the leaps are more elastic and ardent now, crossing the river has become a fun game, a demonstration of skill, strength and balance.

Dance forward, jump backwards, to the sides, with a turn, use your imagination of movement.

Let someone in each group dance ahead, the others following along, imitating simultaneously. The leader uses their whole body vigorously and their whole repertoire of jumping and balancing movements.

Meta-talk on Density

How comes the ice floes float on the water in the river? Even with a duck or a swan on it, or a school kid taking a shortcut across the river.

Discuss *density* related to ice and water. What happens when we put out a bucket full of water in the frosty night? The bucket has cracked by morning and the ice has risen; like a hard hat it extends a bit over the edge of the bucket. The water has frozen and swollen. If we plane off the ice hat, so the volume of ice in the bucket equals the volume of water we poured in—which bucket weighs less? The bucket of ice, as we took away some of the water that had frozen and expanded. If we fill a similar bucket with feathers, which one weighs the least, the one with water, with ice or with the feathers? If we fill the bucket with sand, with dry sand or with moist sand? With lead …?

Now think of a small bucket or a jar that holds one litre, or 1 dm^3. One litre of water weighs 1 kg: the water's *density,* its *volumetric mass density* or *specific mass* is 1 kg/dm^3—when the water is 4 °C (see 'Dancing Water Molecules' in this chapter). The same volume of ice weighs less. The density of ice is lower, why an ice floe floats on water. It sinks a little bit, until the water it presses away weighs the same as its own weight. It sinks a bit more when one of us steps on it, until the water pressed away equals its weight plus ours. We'll deal more with this when we take a bath with Archimedes (see 'Archimedes' Bathtub' in this chapter.)

Floating, Driving, Sinking

Music: sea roar, water rush, pouring creek.

Everyone is spread out in the room. We imagine moving in clean body-temperature water in a little pond or a bay. We float and move light and smoothly, with a feeling of weightlessness, we billow in different directions, around and past each other like small stream whirls.

Let's explore what floats and what sinks in water.

> Imagine you are small feathers that fell into the water. A young bird has lost a few feathers as it flew over the pond. How does a little feather move on the water surface? Someone lies down on their back holding their arms against the body and making small wavy movements. Someone huddles up like a little wad of down. A puff of wind sweeps over the water – what happens? Someone moves sideways on tiptoe holding their body levelled over the floor.
>
> What if you were corks? Some bob up and down, arms by the sides, some with arms crossed. Think of a float on a fishing line, then comes a nibble. Some go straight down. Think of a butterfly's wing, a piece of paper, a mussel shell. A grain of sand blows down in the water. A stone is tossed in. A flat stone is skipped across the water. A boiled egg, an unboiled egg. A dead fish, a sock, a plastic boll, a car tyre, a water-lily leaf, a rowboat, a submarine, an ice floe …

Ask for more ideas. Explore with body and movement whether the water bears or passes through the object. Summarise the different objects' densities compared with the water's solidity.

Life in Creeks, Lakes and Oceans—Mirror Dances

Music: sea noise, rain, rapids, whale song.

Everyone is standing two by two facing each other. One person in the pair is the movement leader, the other one mirrors the leader's movements attentively. The leader starts easily, with small movements allowing the other one to hang in.

> Dance like small fish or big fish, like herrings, pikes, eels, like dolphins, turtles, sharks, whales, alone or in a shoal. Dance like seaweed, algae, corals, anemones, plankton, a jellyfish, a shrimp, a starfish, a seahorse, a mermaid…
>
> Dance peacefully like *amphipods* or *opossum shrimps*. Dance swift like *water spiders*. Dance gracefully like *dragonflies*, or *mayflies* who only get to dance for one day!
>
> Dance quick and energetically like *skippers* on a pond, swish, swish, dodging the other dancers. Dance sideways on all fours, or on all tens, like *green shore crabs*, or backwards like *lobsters*. Gallop through water like peg-like *pipefishes*, small marine animals akin to *seahorses* and *sea dragons*. Dance in circles like *bristle worms*, who eat and do a pooh of sand, like earthworms, though they leave it neatly behind on the beach as small coiling sausages.

The movement leader choses a living organism in the water, a creepy-crawly, a fish or some other animal, a plant or a fantasy figure. Encourage variation of movement, alternating between tranquil and lively, between larger displacements through the water or lingering on one spot.

After a while the other one in the pair takes the lead. Then in the next phase both lead as much as they mirror, carefully giving and taking. After a while two pairs merge. Every person leads as much as they follow impulses from the others. Then two quartets merge and move jointly as the movements evolve. Two quartets join into an octet, and so on, until everybody participates as one entire body of movement. Everyone is alert to the unexpected, which can come from someone else as well as from oneself. Encourage attention, willingness to follow and interplay, a joint imagination of life and movement in water.

> The water temperature goes down. A new ice age arrives. The movements diminish and stiffen. Arrive at a common end.

Turn off the music, sit down, talk about all the animals, plants and creatures they danced.

Meta-talk—Knowing by Naming

Ask for descriptions of movement patterns, appearances, names and properties. Give word support: billowing, jerky, rhythmic, rising, heaving, sinking, awaiting, luring, a shiver, a sudden outburst. A young pupil once showed how an *Gammarus locusta* opens and closes like a soft little jaw. Seek help from wall charts and the Internet.

Make lists of the animals and plants that occurred in the dances. Do they live in lakes, in fresh water, in brackish or saltwater? In creeks or forest meres? Deep down on the bottom? Are there species who come up on land? What are those animals called? *Amphibians*. Which ones are *fish, mammals, reptiles*? Catalogue: fish, cartilaginous fish, molluscs, amphibians, invertebrates, mammals, birds, insects, arachnids, algae, thalloid plants.

List names of *watercourses*: river, stream, brook, creek, mere, marsh, swamp, ocean, sea, lake, pond, canal, ditch. Make lists of sayings: *only a drop in the ocean*; *a drop in the bucket*; *blood is thicker than water*; *thin as reeds in the wind*; *water under the bridge*; *to go into deep water*; *to skate on thin ice*.

What is 'water' in other languages: *water, agua, eau, Wasser, su*, ماء.

Dancing Water Molecules

Music: airy, pouring, e.g. Alan Stivell: Mona.

Everyone is standing in a circle. Assign each in turn a letter: H, O, H, H, O, H, H, O, H, the whole circle around.

> The Os stand straddle-legged, hands on hips and elbows pointing to the sides, making themselves broad.

What does *O* stand for in chemistry language, which *element* does *O* signify? *Oxygen*. From Greek *oxys* and *genēs*: *acid-generating*. Introduce the Os as *oxygen atoms*. What is an *atom*? The smallest particle of a chemical element you can divide the element into, it still being that element. Atom comes from Greek *atomos*, which means *indivisible*. Now we know an atom in turn consists of a *nucleus* of *protons* and *neutrons* around which *electrons* move. We also know even an atom nucleus can be split—we will come back to it later. So an oxygen atom is oxygen's smallest component. If you take away anything more in the atom, an electron or a proton, we no longer deal with oxygen.

The Hs stand with their upper arms up shoulder-high, their lower arms right-angled upward.

What does *H* stand for in chemistry language? *Hydrogen*, from Greek *hydro* and *genēs*: *water generating*. Introduce the Hs as *hydrogen atoms*.

Everybody gathers three by three: each O gets two Hs on either side hooking up their arms in the Os′ arms.

How do we write these compositions of elements in chemistry language? H_2O. We have seen the symbol on signs by a tap or a fire hydrant or on charts.

In each trio the three atoms stick together: two hydrogen atoms and one oxygen atom. What is such a bond of atoms called? A *molecule*. In this case each trio composes a *water molecule*. A molecule is the smallest part you can divide a certain matter into, in this case *water*, it still being water. *Molecule* comes from Neo-Latin *molecula*, which means *little mass*. If we would remove any bit, for example an H-atom from the H_2O-molecule, we would be left with OH, an ion, and we would no longer be dealing with water. A molecule, just those standing here three by three, is a composition or a bond of atoms. What makes them bond so tightly is that the O- and H-atoms are very keen to share their electrons with each other. How?

In science there is always a *how* or a *why*. These are the fundamental questions in all knowing: Why are things as they are? Which conditions are at hand? In the dance following this one we'll go precisely into why one oxygen atom and two hydrogen atoms get hitched so happily and put their electron bags together, and learn why that bond is so strong and why it is called a *polar* bond.

How many water molecules are there actually in one single drop of water? A hundred? A thousand? A million? If a million is a 1 with six zeros—1,000,000—then the answer is a 1 with twenty-one zeros: *1 000 000 000 000 000 000 000*. It's easier to write 10^{21}, ten *raised* to the twenty-first power. Think a billion water molecules a billion times and then thousand times as many again … in one little small waterdrop.

Put the music on.

We imagine a pond. We dance the water molecules in all the drops of water in the pond. Start in a minimalistic way. The atoms sit tightly together one to another, but they can move closer or longer away from each other, thanks to the greater mobility of the electrons. Move about, stretching and contracting, shrinking and expanding, billowing, floating. Pleasantly and peacefully, round the whole pond, between the bottom and the surface.

The water temperature is between some 20° and a comfortable body temperature of 37°. The temperature slowly increases.

The movements increase, a bit more grumbling. The molecules try to get away from each other a little more. Still, the atoms within the molecules hold on to each other. At 60° the movements are larger and more restless, the molecules get further away from each other. Still more at 90°: they root and drag, they thin out and drift longer away from each other.

What happens at 100°? The water *boils* and turns into *steam*.

The dance gets frenetic and chaotic, everyone quivers and bubbles all over the place. Internally the atoms tear at each other and try to drift as far away as possible. At the same time, three by three composing the loose molecules they are, they float and fleet up in the air, light and volatile.

The temperature gradually goes down again, first just under 100°. The steam *condenses* into a *fluid* state again.

The molecules draw closer to each other. 60°, down to 37° again, lukewarm and comfortable, the movements calm down, to 20°, to 15°, 10°, the activity decreases, the atoms and molecules maintain a minimal mobility. We linger a while at 4°. Hold your spatial positions, before we go down to 0°.

What will happen at zero degrees? The water will freeze into ice, taking a *solid form*. When do the molecules get the closest to each other? Think of the bucket filled with water that cracked in the frost night. Or the line pipes in frost. Ice takes more place than fluid water. As we know, ice has a lower density and floats on water. So, at which temperature does water take the least space? When do the molecules and the atoms sit the tightest onto each other? At some degrees above zero: at 4°. That is where water's density is the highest.

On the dance floor the temperature goes down from 4° to 0°.

The molecules shuffle about quietly, shifting their constitution from the highest solidity down to remarkably thinner and finally freezing into stillness.

Zero degrees is the water's *freezing point*. When water freezes to ice the molecules arrange themselves in a hexagonal crystal pattern and take more place. Ice expands one tenth.

One hundred degrees is the water's *boiling point*. The water boils into *steam* and transfers from a *fluid* form into a *gaseous* state. At 100° the steam is moist, as it is still mixed with fluid. As the temperature continues to rise the steam finally becomes *dry* or *satiated* and everything turns into gas. But the atoms hold on to each other at all temperatures, through thick and thin—they fancy each other, devotedly shaping the water molecules.

Atoms Flirting and Bonding into Water Molecules

Music: rhythmic duple or quadruple, e.g. Radio Tarifa: Rumba Argelina; see detailed description of basic steps in 'Galloping or Skipping Two by Two' in Chap. 2.

Everyone is standing in two lines beside each other, one line at each short end of the room. Show the basic steps with one of the dancers:

One dancer from each side skips forward, hands at their sides. They meet in the middle, grab each other arm-in-arm, make a few turns around, then change sides. Everyone, in turn, practices the steps two by two to the music. The others clap the beat. When everyone has danced two at the time the two lines dance towards the middle, wave and say bye, bye, and skip backwards, hands on hips.

Introduce the water molecule theme.

> In one line everyone is standing like oxygen atoms, steady Os, hands on hips. In the opposite line everyone is standing ready like hydrogen atoms, arms shaping an H. To the music an oxygen atom skips forward to the middle, dancing and showing off, spinning round, turning one shoulder and then the other shoulder to the H-atoms, glaring at them, tempting and flirting. Eventually two H-atoms bite and come dancing. They hook on one at each side of the O-atom, forming a trio. The three dance round a few happy turns, then they skip away and live happily the rest of their days in the eternal cycle we deal with later on.
>
> A new O-atom dances forward, attracts two new H-atoms, they merge and dance along and away.

The dance continues until everyone has danced forward and flirted or been asked for a dance. Fill in with hydrogen and oxygen atoms if needed. Switch, so everyone gets to dance like oxygen or hydrogen atoms building water molecules together.

Meta-talk on the Chemistry Between the Atoms in an H_2O-Molecule

How come two hydrogen atoms and one oxygen atom are so attractive to each other? Visualise this with the Rutherford-Bohr atom model, on the board or with three dancers, some larger circles and a few small balls. An oxygen atom has six electrons in its exterior L-shell, which is two short of a stabile set of eight. A hydrogen atom has only one electron on its K-shell, one short of a stabile set of two. Sharing their assets, the O-atom gains one electron from each of the H-atoms, who in turn can 'hang' with one of the O-atom's six electrons in its outer shell. The three become bosom buddies and everyone is happy and contented.

The water molecule's bond is called *polar*. It is stable and content in itself. It holds water together in drops when contacting other material: a pane of glass, a marble slab or a woollen sweater, and similarly sucks up salt or sugar easily. It allows striders to skate around on the water surface on their three pairs of hair piles—hairy legs—without sinking down, thanks to the *surface tension*. Also the *capillary force* allowing sap to draw up in plants and trees in springtime can be explained by the water molecules' strong, polar bond.

Water Sports—Slalom

Music: rhythmic, driving triple beat, e.g. Shchedrin, Carmen Suite: overture + ballet.

Half the group is sitting cross-legged behind each other in a line in the middle of the floor, leaving enough space to allow the dancers to move slalom between them. Those sitting clap the triple time in the music as the skiers take off: the other half, standing in line in front of them, pepping and getting ready for the downhill run.

Show a sweeping, rocking slalom dance movement. Arms parallel, pendulating intensively from side to side, shoulders and upper body rocking along asymmetrically in the opposite direction, feet tripping frenetically on tiptoe.

One after the other the slalom dancers dash on in a fast and winding line between the gates, arms swinging in and out in every passage, waists twisting in the opposite direction, inward in front of the next gate. Every odd dancer sways right between two gates, every even dancer goes left, a coordinated crossing movement. After the last gate they rush back along the sides to the top and ski down the slope in a new lap between the gates, then another one and another one and another.

Accelerate on to a downhill race, a steeper slope, the racers huddling up, arching their backs, elbows tight against their bodies, fists pressed to their breasts, clinched around the ski poles, the curves getting tighter. A few dashing laps one after the other.

A couple of laps later the skiers start going slower and slower as they get more and more tired, their legs get unstable, they start stumbling along, having greater and greater difficulty keeping their balance and measuring their way through the gates. They orbit far out to the sides, arms fluttering hither and thither, trying to maintain the pendulating movement, getting all pumped out and loose-limbed. Ultimately they crash and slump down on the ski slope.

Choreographing More Water Sports

Ask for suggestions on other water-based sports: diving, curling, giant slalom, synchronised swimming, water polo, ice-dancing, canoeing, surfing, windsurfing.

Ice hockey. Ask for suggestions of a typical movement. Skating on with the puck by the stick. Stylise, rhythmise in common time, use the whole body and the whole floor area. Refine and rehearse the movement together. Call it the first. Ask for a second movement, a third, a fourth, a fifth. For example: facing off, taking a shot at the goal, parrying a shot as goaltender, skating backwards, being sent to the penalty box and sitting down all in a row on the bench, tapping the sticks on the floor to pep up the team on the ice. Proceed the same way: stylise, put it into a quadruple time, rehearse. Call the movements second, third, fourth and fifth. Rehearse the whole sequence a few times, four beats at the time.

Put it together into a dance. Put the music on, lead the dance or let someone lead it by calling out the numbers, first from first to fifth, then in a random order, four beats for each movement.

Archimedes' Bathtub

This is a concentration exercise. Everybody stands spread out in the room, eyes closed. We imagine we've run a bathtub full of body-temperature water. The tub is filled to the brim, no space for one more drop. We are about to step into the tub, longing for a nice and relaxing bath.

Lift a leg. Slowly, carefully dip a toe in the water. The tub is filled to the brim. What happens to the water as you dip in your toe? It runs over the edge. How much water runs over? A toe's volume, as much water as your toe presses over. Go on and insert your whole foot into det lukewarm water. How much water runs over now? A 'foot' of water. Sink down your foot onto the bottom of the bathtub, your leg going down to your knee. How much water runs out over the edge? A reciprocal volume to what the lower leg presses out. Take down the other leg, toe first, then foot, calf, up to your knee, more water runs over. Crouch down, your thighs go down, buttocks, your stomach. Hold on to the edge of the tub, sit down. Lots of water run out over the edge. Put your hands on the bottom, let the weight of your body come down, lean back, your torso goes down, your shoulders. Let your neck go down, your chin, your mouth. Hold your nose, dip down your head. Now your whole body is under the water, quiet and embraced by lukewarm water. Relax and enjoy the moment.

Lift your head out of the water, lean your back on to the tub. Open your eyes.

How much water has run over the edge of the bathtub altogether?

Meta-talk on Archimedes' Principle

Which physical law becomes manifest here? We've all noticed how our bodies become lighter when we lie in a bathtub or dive down into a swimming pool. You can almost lift a friend or your dad playing in the lake or the ocean. You can lie and float on the water.

Just how much lighter do you actually get? Is there a way to measure this? How much is the weight reduction? It is equal to the weight of the water that runs over the edge of the tub. Equal to the weight of the water the body presses out over the edge. Which is as much as the body's volume in water. Or as much as we would weigh if our body consisted of water.

Who had this flash of genius? Archimedes, the good old Greek! As he was sitting in his tub enjoying a bath, just like we do now, sitting and thinking about it too. Suddenly it struck him and he jumped out of his tub and ran out of his house shouting: 'Eureka!'—I've found it! He was running and shouting stark naked, according to the legend, through the streets of his hometown, the harbour of Syracuse in Sicily.

Follow John / Yoanna

Everyone lines up behind each other. The one in front chooses trajectory and movements, inspired by the water theme: an animal, a fish, a plant, the water itself, running, purling, rushing or pouring in waves, pelting rain, hail. Everybody follows the leader and imitates their movements. After a while someone else hops in front and takes the lead, then someone new, and another one, quick changes until everyone has led with water-inspired movements.

Water in the Body, on the Earth, in Fruit and Vegetables

Make the circle with the H_2O-molecules again. How many water molecules were there in a drop of water? 1 000 000 000 000 000 000 000, or 1×10^{21}.

In the circle the dancers stand three by three. In a water molecule there are two hydrogen atoms for every oxygen atom. How big a part of the human body is made up of water? Pretty much like we stand here three by three: a good two thirds of our body is water. The rest is mostly oxygen and hydrogen, besides salts, carbon, nitrogen, potassium, phosphorous and iron. The muscles contain two thirds water, the skeleton almost none. Therefore our body needs a lot of water, one to two litres a day.

How much is in fruit and vegetables water? Cucumber contains 97 percent water, like watermelon, which is a *cucurbit* as well, as are pumpkin and squash. Animals also consist largely of water. The most watery are jellyfish: up to 98 percent.

How much of the earth's surface is covered by water? Here too the ratio is a good two thirds of water, of which 94 percent salt and 6 percent sweet water. Water is a precondition to life on our planet. All life originates from the sea. Now we are looking for traces of oceans and water on Mars and other places in space—is there life somewhere else? Are we alone in the universe? We will return to that.

Body Fluids

Music: 4 different tunes, from steady rhythm to rattling to sneaking, e.g. Radio Tarifa: Las Cuevas, Conductus, Baile de Almut, Canción Sefardí; see detailed description of the basic steps in 'Sprinters, Sneakers, Stressers and Slackers' in Chap. 4.

Which fluids are there in the human body? Blood, saliva, sweat, tears, pee, pus, slime. Ask the group to divide up, equal numbers in each corner.

Introduce the first gang: the *pee people*.

They run around light and jolly, with *the whole sole of the foot* on the floor. They run forward, backwards, to the sides, round, round. Like pouring water. Sometimes they hop up jumping for joy, they splash and spatter, rushing on like streaming water, touching the floor with the whole sole of the foot.

Put on the music, the pee people dance all over the floor. Remind them about the contact of the whole sole of their feet on the floor, and to jump for joy now and then. After a while they return to their corner—applause for the pee people!

Introduce the second bunch: the *wet breath people*.

Breathe against a window pane, it gets misted over. There is much water in breath.

These people steal around invisibly, they sneak *on tiptoe*, forward, backwards, to the side and round, round, round. Their bodies are all light and smooth, they move light and airy, like the steam in the sauna or the mist over a summer meadow at dawn or dusk. Now and again they quickly squat down, they do not want to be seen nor heard, they rise slowly, then fume on, invisibly and soundless, all the while on tiptoe.

Put on the music. The wet breath people spread out in the room, sneaking everywhere, between the walls and the ceiling, into the smallest corners. Remind them to stay on tiptoe the whole time, to stretch out their arms and legs to all sides filling the whole space, and to squat down and make themselves invisible now and then. After a while they sneak back home to their corner—applause for the wet breath people!

Introduce the third gang: the *sweat people.*

These people work frenetically, they are always in a rush, getting all sweaty. They don't have time to put down their whole feet, they stress along on their *heels* all the time, forward, backwards, to the sides, round, round. They move about like small bouncing bolls, they labour and shake with their whole bodies, legs, arms, elbows, heads, hair. Time again they check their clocks, jumping up with high knees, stressing on and sweating even more.

Put on the music, the sweating people start tripping and stressing about for a while. After a hard-working day and without slowing down, they finally stress their way home. Applause for the sweating people! You get all sweaty just watching them.

Introduce the fourth gang: the *snot-and-slime people.*

These people hardly manage to shuffle about at all, moving *on the inside and the outside of their feet*, forward, backwards, to the side, round, round without purpose or direction, heavy and sticky. Their bodies are feeble, as in a state of dissolution. Time after time they collapse, all weary and worn out, a formless amoeba draining over the floor like a wet spot. But they too must carry on. They drag themselves up with big effort, and dawdle and loiter on, on the inside or the outside of their feet.

Put the music on, the snot-and-slime people stagger and drag themselves around as well as they can. Remind them about their feet's contact with the floor. Eventually these folks trudge home. A round of applause for the snot-and-slime people, who did the best they could.

When every group has danced in their way, out of their corners and back, receiving well-earned applause, everybody gets to dance in turn as pee people, wet-breath people, sweat people and snot-and-slime people.

A Glass of Water

In this concentration exercise, which a few thousands of students have done over the years, the glass of water is imaginary. Through imagination we can nonetheless evoke a sensory perception of the water's eternal cycle, of which humans are a part. It has been going on ever since the first drops of water were formed on our planet. Humanity's impact on the planet evolves with alarming disturbances pressing the whole ecological system towards an uncertain future.

At the Nobel week's dialogue day in Stockholm, Sweden, in December 2018, artist Marina Abramović staged a performance on the theme of climate threat similar to this concentration exercise. Everyone in the aula received a glass of water. Abramović asked everyone to close their eyes, to slowly move the glass to their lips, sense it and take a sip. Here, we imagine through fantasy a glass of water in a concentrated bodily, kinaesthetic experience.

Everyone is sitting spread out over the floor.

> Close your eyes. Imagine it is high summer, the hottest day of the year. You are sitting on a sandy beach, you have taken a swim and the heat of the sun has dried every drop of water on your skin in a jiffy. Your lips and mouth are all dry, you yearn for a glass of water, clear and cool.
>
> Imagine someone hands you a glass of water. Take the glass, feel the coolness against your fingertips, how it spreads through your fingers, your hand, all your senses. Slowly move the glass towards your lips, longing for a drink. Touch the glass and the water with your lips, with the tip of your tongue, take a tiny sip.
>
> Take a little gulp, feel the clear freshness of the water against and inside your lips, against your teeth, under and over your tongue, against your gums, inside your whole mouth.
>
> Swallow the gulp, take another one. The water runs down your throat, refreshing, down your gullet, feel it inside your chest, down into your stomach, while the sun continues to warm your skin.
>
> Empty the whole glass, slake your thirst.

Now think about the water you just drank from the glass. Think, that in those gulps of water there may be a drop that a dinosaur drank two hundred million years ago in a river here on earth. Or a drop that Alexander the Great sweated out during his campaign against the Persians three hundred years before the common era. Or a drop that a tiger in a jungle deep in Africa peed out a hundred years ago.

Some of the drops of water that ran into your stomach and which you will sweat or pee out, maybe will, a couple of centuries from now, be drunk by a distant relative pumping it up in the clear and purified water from a suspended well. Or by a tiger in Africa's jungle, if there still are tigers in Africa, if jungles still exist, if we take care of our planet.

Discuss the eternal cycle. Talk about how we handle water. Access to water in our homes has not been self-evident during the latest hot summers. In many places around the world water is more than ever before a commodity in short supply, soon a commercial commodity, something to migrate to or wage war about. Discuss our responsibility for water and the planet.

Water Cycle Dances in Canon

(by Stefan Forssén, see https://ejeby.see/produkt/water canon/)

> 4/4 //:Water clouds keep raining down
> as rain or snow are falling down
> pouring on in purling creeks,
> sliding into floods and streams,
> stalling in some peaceful lake
> or floating out into the sea.
> Sunrays warm up lakes and seas,
> damp and moist rise to the skies
> building://

Rehearse the song, try to sing in canon. Divide the group in two. Each half puts on a dance to the song.

Encourage movement variation in space, tempo, with the whole body or single body parts, shifting between forceful or holding back, between moving around and standing, sitting or lying down, doing some movements jointly, others with only a few, making a circle, or several, dancing a sequence in a loop through the room, and so on. Encourage free imagination of movement with respect to the lyrics. The emphasis does not need to be on illustrating or dramatising the song, nor on simultaneity with the lyrics. Focus rather on a feeling of how clouds move, how snowflakes fall, how water gushes, a corporeal experience of the windless calm on a lake, stormy ocean waves or a wet airstream rising to the skies. Try the feeling of dancing one single quality of movement through the whole song.

One half of the group dances to the other one singing. First in a slow tempo, then a repeat a little faster, then again either in an even faster pace or in slow motion.

Waves Cradling the Winds to Rest

Music: sea rush, waves, e.g. Radio Tarifa: Soledad; prop: big fan; see description of the basic movements in 'Rocking and Rolling Waves' in Chap. 2.

> Everyone lies on their back, spread over the floor, eyes closed, recovering their breath. Fantasise we are lying on a sandy beach by the sea. A light breeze fans over the sand from the water, rippling its surface.

Go round and fan gently over the lying dancers with a big fan.

> Fantasise we are the water, rippling in the wind. The wind increases, the waves grow bigger, rocking from side to side. Rock your bodies gently from side to side. The wind gets still stronger, the waves roll over, back and forth in the wind.

Mark the wave movements with the fan and your voice:

> Roll over, roll back, and forth … and back, and forth … and back, the eternal swell of the waves on the beach..

Let the wind ease down with the fan.

> The winds abate, the waves come to rest.

Read Indra's daughter in Strindberg's *A Dreamplay*, standing on the beach and listening to the waves singing:

> We, we, waves,
> are rocking the winds
> to rest!
>
> Green cradles, we waves
> Wet we are, and salty;
> like flames of fire;
> wet flames we are.

Dousing, burning,
washing, bathing,
breeding, bearing.

We, we, waves,
are rocking the winds
to rest!

Stay on the floor lying down another while. Think about it: we, we waves rock the winds to rest. Think about everything we have danced and played today, and everything we have learned about water, while dancing water.

Air

Put very simply:

Air is what we breathe …

Yoga Warm-up Taking Deep Breaths

Everyone stands in a big circle. Lead a warm-up with inspiration from the yoga positions *sun greeting* and *downward dog*. Do the sequence three times right and left.

Stand with a straight spine all the way up to the neck. Feet are shoulder width apart for stable contact with the floor, so the whole body weight from crown to sole rests steadily on heels, balls of the foot and toes. Arms are hanging relaxed by the body. Take deep slow breaths and move into the following positions in an uninterrupted flow, your joints and muscles stretched a little more with every repetition.

Turn palms forward, breathe in, lift arms straight out and up in a half circle over the head, look up to the sun, put palms together – the *sun greeting*. Breathe out, lower hands in front of your face all the way down, bow your neck, your back, bend your knees, put your palms on the floor in front of your feet. Breathe in, stretch one leg all the way back, toes touching the floor, look up to the sky, breathe out. Breathe in, look down, stretch the other leg back, toes on the floor, curve your back, stretch the back of both legs as much as you can, press down your heels, look at your feet – the *down dog* position – breathe out. Breathe in, bend your arms, dive forward with your shoulders and upper body, chest almost touching the floor, look up to the sky – the *cobra* position – breathe out, curve your back inwards forcefully down, pelvis to the floor, chest pressing forward, straightening up arms. Breathe in, arch back, back to the *down dog*, look at your feet, legs as straight as you can, press heels down to the floor, breathe out. Breathe in, pull the other leg back in, foot to the starting position, look up to the sky, breathe out. Breathe in, pull the first leg back in, feet back together, breathe out. Breathe in, rise, straighten your legs, roll up: back, neck, back of the head, palms forward, arms out and up like in the beginning, look up to the sun, start all over again from the sun greeting.

Do the whole flow of movements again, first one foot backwards, then once with the first foot and 1 s time with the other foot. Take deep breaths, in through your nose, out through your mouth. Do the yoga movements at a slow pace, fully kinaesthetically aware, stretching muscles well. Later we will dance the oxygen intake thanks to the inhalation, making the muscles grow and the body survive, and allowing us to move at all.

We can also put it this way:

Air is what birds fly through ...

The Birds' Dance

Music: free flow with rhythmical parts, e.g. Radio Tarifa: Canción Sefardí; see detailed description in 'Arms Like Wings to Fly With' in Chap. 2.

Crouch down. Imagine we are birds, our arms are wings. It's dark, the birds are sitting in the tree, sleeping. Morning hasn't broken yet, it's cold, we keep our wings folded on our backs. Hear the music, here comes the sun.

The sun warms the earth and the air around the tree's branches and leaves. The birds wake up, longing to get out and sail in the upward rising air. Unfold your wings a little and feel if the air is warm yet ... brr ... doubt it ... ruffle up your feathers, wait a while. Time to get up? No, brr ... what a chilly morning!

Let's wait a little longer. Feel again ... Yes, now the air feels good. Stretch out, unfold your wings, make big strong wing strokes to warm up. Feet move quickly but quietly, tripping on tiptoe, the bird's dance is a soundless dance.

Up, up and away, fly, crisscross. Try your wings, strong wings growing stronger. Let an air spiral take you upwards. Spin round, wings stretched out to the sides, feel the air underneath. Spin round in one direction, in the other direction. Stop, train wing beats, listen to the rhythm: lift your wings out and up high over your head, lower your body, bend your knees. Press your wings down and stretch your body on straight legs, up on your toes. Lift your wings and go down through your knees again, squat down, press your wings down again, heave your body up to the sky again, to the rhythm of the music, wings up, down, up, down, the body down and up, down and up.

Fly through the skies again, forward, backwards, to the sides, spin round. Point one wing down to the ground, the other up to the sky, look down on earth. Change, point the other wing up to the clouds, look up, turn. It may make you get giddy. Fly away with pace, in different directions, break in the air, go on in some direction, in circles, glide, full-fledged air acrobats never clashing into each other.

Time to fly home to the tree, dusk is setting in. Land on a branch, wings risen, crouch down, wings down. Jump up, wings up, change branch, wings down. Change branch again, again, and again, one last time.

Land for the night. Fold your wings, crouch down, spread out your wings and shake your feathers, fold them behind your back. What a day for flying it has been, ruffle up, go to sleep.

Meta-talk on Dance, Physics and Biology

Talk about the dance, how it feels in your body, how your arms got to work hard, shoulders too, and thighs. How you can move and dance in so many different ways, and get stronger all the time, in your muscles, in every manner. How do birds dance?

Move the discussion on to talk about birds' movements, how they interplay with the air. How do the birds manage to keep themselves up in the sky, without crashing to the ground? Don't they weigh anything? Discuss birds' adapted weight, their skeleton's hollow bones, their almost weightless feathers, which they can spread for maximum contact area with the air. We pressed our wings down, pressing the air underneath together and heaving us up, like on an air cushion.

Can air be pressed together? Air is matter, as we'll see in the next dance. When we press the air together with our wings, we create an overpressure, while above our wings, where we pushed away the air, an underpressure is created. How does an airplane stay up in the air, they don't flap their wings? As the aircraft gains speed on the runway, an overpressure is created under the slightly angled wings, and an underpressure over the wings, so the aircraft is sucked up into the air. In Arabic airplane is طائرة (*taa'irah*), feminine of bird طير (*taiir*)—who determined such a thing, Palestine poet Mahmoud Darwish once wondered. What are bird and airplane in other mother tongues?

Swallows Sailing Under the Clouds

Music: free flow, e.g. Gipsy Kings: Passion; props: a long piece of fabric or sheeting, preferably light blue with white spots, lying across the middle of the floor.

Half the group is standing in a row beside each other at one side of the room, the other half on the other side. Each group picks a species of swallow to dance: *barn swallow, cliff swallow, bank swallow.*

Hold one end of the fabric, one of the dancers holds the other end, holding the ends a bit apart. Lift it together so it heaves, and let it slowly sink down, pull it up again and let it fall down, at the rhythm of the music, like surging veils of clouds for the swallows to fly under. The swallows may not brush against the clouds, so their feathers don't get wet and they crash to the ground. They sail soundless under the cloud, wings spread out.

Put the music on. Heave the fabric so it bellows in an unceasing movement up and down.

> Fly, barn swallows! The one group take off spreading their wings, glide under the cloud and on to the other side and then back again, softly swishing, soundless. Once again! They time the fabric's swaying so they just manage to pass underneath. Keep your wings spread!
>
> Fly, cliff swallows! The other group lift their wings and come gliding, and then back. Once more! Now all the swallows fly ahead! They all sail on and pass each other under the cloud. Fly back, all you swallows! One more time, without clashing!

Keep the cloud's movement going.

Fly, barn swallows, fly! Come, land and rest under the cloud!

The fabric falls softly to the floor, covering the dancers, then lifts again.

Fly away barn swallows! Fly, cliff swallows, fly! Come, land and lie down to rest under the cloud!

The cliff swallows come flying and time their landing under the fabric, rest there a while until the cloud bellows up again and then fly away.

Come fly, all you swallows, and lie down to rest under the cloud!

The fabric sinks and covers all the swallows, hiding them out of sight.

Meta-talk on Swallows, the Air's Matter, Mass and Weight

Talk about the colour and appearance of swallows, their slender bodies and pointed wings and tails. Which ones have white bellies, black, dark blue or brownish backs, red throats, fork-tails or square tails? *Swifts* with their black-brown backs and bellies look similar, but they are not swallows. What do *paper swallows* look like, how do they make their way through the air? Discuss where swallows live in summer and in winter, how they throw themselves out of their nests, fly high up to the clouds, over acres, meadows or steeps, and always return to their nests when spring comes. Why do swallows at times sail high up in the skies in summer and at times swish low over the ground? What are they looking for? At low pressure the swallows fly low, at high pressure they fly high. Why? They fly where their food is—mosquitos, insects!

Why does the cloud fabric in the dance behave like it does, slowly up and down, softly billowing? When we lift the fabric, we need to put some force to it. It shapes like a tub or a hammock. The air shapes it. Air is a gas surrounding us. Like steam, air is matter. It consists of particles, mostly nitrogen and oxygen, plus noble metals and carbon dioxide and a little hydrogen. Air has mass and is drawn to the earth by gravity: it has weight.

The air pays resistance to the fabric's movement through its weight. The air mass which the fabric cups around weighs more than the fabric itself.

When it falls down it does so slowly, shaping an arch over the air caught beneath it. The air bears it for a while, is pressed away bit by bit and the cloud fabric sinks to the floor.

We can think of the pillar of air every one of us has on top of our head. It reaches up to the end of the atmosphere, a hundred kilometres high and it weighs over three hundred kilogrammes …

Finally, we can also say:

Air becomes winds when set in motion.

Flag Dance—Winds, Thermals, Gravitation and the Earth's Rotation

Music: duple or quadruple, driving, e.g. Gipsy Kings: Viento del Arena; props: flag, angle chimes, fans, widening plastic balls.

Everyone dances in a line skipping behind each other, a loop winding through the room. The leader holds a big, colourful flag with a thin fabric, everyone else skips behind hands on their backs, knees high. Beat the rhythm on the drum or tambourine. Stop the loop now and then and put in an extra movement on the spot, a turn, a spiral, a whirl. At the end the leader leads the gang into a circle round, round in the middle, side-galloping.

Meta-talk on Winds—Air in Movement

Why does the flag fold out and flap in the dance? It forces its way through the air as the flag holder moves through the room, through the air all around us, filling every tiny space. Air is matter and has mass and weight, as the swallows' dance under the cloud showed us. The air resists, as manifested by the cloud fabric. You see it and feel it in your hair and against your cheeks dancing behind the flag and side-galloping in the circle. The air is pushed aside and set in motion.

Why does a flag on a flagpole float? No one is holding it or dancing around with it? The *wind!* The wind is blowing, the air has come to move. What are winds? Air in motion. How does the movement in the air start, how do winds arise? Discuss the three main forces: thermals—that is, the difference between warm and cold air—gravity and the earth's rotation.

Thermals: As the sun warms up the air, it ascends. Colder air fills up underneath, warms up and rises. Air typically makes the winds blow in a spiral movement. Light the candles in an angel chime. Discuss the air movements involved, visible and invisible. The thermal spiral movement is visible also in the autumn leaves whirling in the school yard.

Gravity: We know that air, like steam, is a gas and consists of small particles. The air has weight, we saw it in the swallow dance: it took time for the cloud fabric to sink to the floor. The air captured underneath the cloth carries the cloth, it weighs almost as much as the fabric before it gradually evacuates. We feel the wind's force and weight in our faces as we dance through the room, or walk into the breeze on the beach or along the facade of a high-rise building.

The earth's rotation: This brings down the air's movement, in a dynamic known as the *Coriolis force*. It works against the movement caused by the thermals and tries to drive the air back to where it blows from. It makes the big winds and ocean currents turn *anticlockwise* in the northern hemisphere, *clockwise* in the southern hemisphere. What about the water in a wash basin or a bathtub when emptied, does it whirl down the valve in different directions in the northern and southern hemisphere? Do the leaves in the schoolyard whirl in different directions in Canada and

Chile? No, those amounts of water or air puffs are way too small to be affected by the Coriolis force. But the big winds and the oceans are affected by the earth's rotation and the Coriolis force.

Sound

The Flag and Sound Stick Dance

Music: duple or quadruple, driving, e.g. Juan Peña Lebrijano & Orquesta Andalusí de Tanger: El Anillo; see detailed description in 'Dancing into the Room in One or Several Loops' in Chap. 2.

The group divides into two lines dancing like in the previous dance. The leader of one line holds a flag and skips ahead through the room, everyone behind skipping along, hands on their backs. The leader of the other line holds a *sound stick*, a stick with bottle caps, jingle bells or pieces of metal nailed onto it, and gallops ahead, the others galloping behind, hands on their backs. At the end of the dance the one group forms an inner circle side-galloping in one direction, while the other group dances around them in an outer circle side-galloping in the other direction.

Meta-talk on Sound—Air Particles in Motion

Discuss the difference in what the senses perceive when the flag floats gently and when it is swayed heavily through the air. It *sounds!* How does sound arise? When air is set in motion, the air particles start trembling and vibrating. The air molecules are jammed together, shoving against each other.

Blow up a paper back and bang it. The air pressure in the bag increases as we blow more air into it. The air thickens and when we smash the bag the air shoots out through a hole and into the room, where the air is thinner and the air pressure lower. When we prick a hole in a filled balloon all the air inside wants to evacuate fast through that little hole, instantly widening it and producing a bang.

Meta-talk on Speech—Voiced and Unvoiced Consonants

The most common sounds, besides our heartbeat, are breathing and speech. Make air currents with your breath: *f f f*. Where does the sound arise? The air puffs rub against the lips. Do *s s s*, *p p p*, *t t t*, *k k k*.

Use the same force blowing and shift between *f f f* and *v v v*, between *t t t* and *d d d*, between *p p p* and *b b b*. Feel the difference in your chest and Adam's apple.

Which consonants engage the vocal cords? *Voiced* or *sounding* consonants. The *unvoiced* or *soundless* do not make the Adam's apple and the vocal cords vibrate.

What does the word *consonant* mean? *Co*-sounding. Consonants sound along with the vocals that sound more. To sound is *sonar* in Spanish, *sonner* in French. What is *resonance*? *Re*-sound, *echo*. The guitar body is a sounding box where the sound from the trembling strings sounds and re-sounds and is amplified and reinforced. Like the echo in a cage or a canyon. All cavities in the body, mouth, sinuses, stomach and chest help amplify the sounds we shape with our mouth and lips as we speak and sing.

Hand Claps—Sordas and Claras

Clap a common time, accent on the beat of one. Clap with cupped hands, the claps sound muffled, *sordas* in Spanish. Clap three fingers kept together hitting the open palm. The claps sound clear, *claras* in Spanish. Some flamenco rhythms to clap and stamp are given below. The claps cause air displacements. Use the same muscle force and shift between *sordas* and *claras*. Why do the claps with the open hands sound more than with cupped hands? The air moves away faster. Some air stays behind, caught between the cupped hands, while all the air blasts away as the fingers hit the palm of the hand. Think of the sound of a match against the rubbing surface on a match box. Or fireworks, or gasoline ignited and exploding, trigging the valves in all cylinders.

Waves of Sound

Take up a big metal spiral, toss it over the floor holding the one end, someone catches the other end and holds it firmly. Make shock waves in the spiral's lengthy direction, see how they bounce back and forth between you and the person on the other end. How does sound travel?

Like *shock waves*: waves of air particles bounce against each other, shoving each other so the air thickens and thins. Waves spread in all directions, like circles on the water in a pond when we throw a stone into it.

Make sideways wave movements over the floor with the spiral, the person keeping the other end still. What is *frequency*? The number of swings per time unit. Make huge, slow swings as to visualise low frequency sound, low tones. Increase the number of swings towards higher tones.

Conclude how sound arises beating the drum. Or when the bottle caps on the sound stick and the bells on the tambourine hit each other during the dance. When the balloon cracks. How fast does sound travel? Through air sound waves travel 340 m per second. Sound waves can also move through fluids, like water, or through a wooden or concrete wall, especially base tones. Can sound travel through a vacuum? What is a vacuum? *Airless space, void!* A space *empty of air*, where there are no air particles that can bounce and shove each other ahead. Airless space is soundless space.

How do we receive sound waves? How does hearing work? We will deal with that later.

Flamenco—Clapping and Stamping

Have shoes, preferably with a small heel and leather soles for a better sound. Stand straight, feet together, arms stretched in front like two brackets, hands together almost at shoulder height. Clap a common time with different accents, alternating *claras* and *sordas*, clear and dull claps. Stamp common time with the whole foot touching the ground (*golpe*), the ball of the foot (*planta*), the tip of the toe (*punta*) or the heel (*tacón*).

Establish a flamenco rhythm called *tangos*, a musical phrase in a 4/4-beat: *one* 2 3 4, *one* 2 3 4, *one* 2 *three* 4, *one* 2 3 4, two times in a row. Clap the phrase twice, everyone listens and then repeats by ear. Count and clap once for those who want to count. For the youngest you can 'count' like this:

> *Long* long long long, *long* long long long, *short* short *short* short,
> *long* long long long.

Rehearse the phrase until it 'settles'.

Now stamp the phrase two times, once with the right/one foot, once with the left/other foot. Hands on hips. Stamp with the whole foot, *golpe*, one stamp for every accent, Rhythmise your voice:

> *right* pam pam pam, *left* pam pam pam, *right* pam *left* pam, *right* pam pam pam, continue
> *left* pam pam pam, *right* pam pam pam, *left* pam *right* pam, *left* pam pam pam.

> *or*: *pam* pam pam pam, *pam* pam pam pam, *pam* pam *pam* pam, *pam* pam pam pam, once more.

For the younger ones:

> *one* foot pam pam, *other* foot pam pam, *one* foot *other* foot, *one* foot pam pam, and again.

On the 'long' stamps we have time to shoot the tip of the toe of the not-stamping foot forward, about a foot's length, the heel lifted a bit up off the floor. As if the stamping foot sends the other foot ahead with a smash. Point with a stretched foot, slightly turned out, flamenco style. On the 'short' stamps both feet stay parallel beside each other.

Do two phrases, start with the right foot, then the left.

Repeat, now with a double stamp: a smaller, marked stamp just before the accentuated stamp, on the upbeat, on the *and* or the *pa*:

> and *one* two three four, and *one* two three four, and *one* two and *three* four, and *one* two three four.

pa-*pam* pam pam pam, pa-*pam* pam pam pam, pa-*pam*, pa-*pam*,
pa-*pam* pam pam pam.

Shoot the not-stamping foot forward again on the 'long' stamps. Do the phrase two times, right foot, left foot. Try a third rhythmic variation: two small stamps on the upbeat with the *first* foot, the other foot stamps the rhythmic phrase, shooting a foot length forward:

pa-pa-*pam* pam pam pam, pa-pa-*pam* pam pam pam, pa-pa-*pam*, pa-pa-*pam*, pa-pa-*pam* pam pam pam – two times, right, left.

Rehearse a sequency with the three variations on the 4/4-phrase, right and left: with a single stamp on the accents, with a little stamp on the upbeat, and with a double stamp on the upbeat and the accentuated stamping foot a bit forward.

The flamenco *tangos* rhythm, 'closes' on the last accent, the last *one*:

pam pam pam pam, *pam* pam pam pam, *pam* pam *pam* pam, *pam!* ... (three silent beats on the last two three four).

Divide the group in two halves. The one half claps, the other half stamps. The accompanists, *palmeras*, clap their hands, two phrases for an intro, setting the rhythmic phrase and the tempo. The dancers, *bailaores*, clap along the two intro phrases, then put their hands on their hips and stamp the whole *tangos* phrase. The *palmeras* and the *bailaores* all finish on the last beat, on the *one*.

Dancing Sound Waves from Drum to Eardrum

Music: 6/4 time with the accent on 1, e.g. Juan Peña Lebrijano & Orquesta Andalusí de Tanger: Del Placer Que Irrita.

Everyone stands in a big circle with a conga drum or a loudspeaker in the centre. In a while the dancers will move into three smaller circles with their backs to the drum or the speaker and dance like sound waves in canon. Show the five different steps first, four measures each.

The rhythm goes: *thump* two three four five six, *thump* two three four five six, *thump* two three *thump* five six, *thump* two three four five six, or:

1 2 3 4 5 6, *1* 2 3 4 5 6, *1* 2 3 *4* 5 6, *1* 2 3 4 5 6.

The first step, with an *accent* on the 1, is set going as if caused by a drum beat, a heartbeat and a thump in the back:

Shoot your chest forward, the upper body billowing forward, skip ahead on your right foot as to regain balance, gather your feet together – all in one move. A new thump on the next 1, a new nudge in the back, chest shoots forward, skip ahead on the same foot, gather feet. The third and fourth thumps come right after each other, the fifth thump is like the first and second.

Do the five thumps and skips in an organic, swaying flow, a wave movement triggered by a mild but distinct shove in the back. The movement impulse is marked by the chest shooting forward, the upper body following behind, the leg skipping ahead in order not to tip, the upper body swaying back; then a new thump, a new shove, five in all.

The second step: on the pulse beat, the right arm extends to the side, then the left, right-left, right:

> Now the wrist marks the *accent* and gives the impulse to the arm movements. The wrist springs out to the side, the hand follows along, the whole arm swings out and up to shoulder height, like a wing beat, falls slowly down, the wrist leading the downward movement. Just when the hand lands back on the side of the body, the next pulse beat comes. The left wrist swings up, same arm movement out and up, the arm falls back, wrist leading. Same again right and left, now twice as fast following the rhythm, and then one more time starting with the right wrist like in the beginning.

The third movement in the choreography is the same as the second movement, now the reciprocal leg and foot extending to the side on the pulse beats as the arm extends out and up. First right, then left, right-left, right.

The fourth movement is the same as the third, now with a sideways displacement in the circle. In the same way as skipping forward in the first movement and gathering the feet together, now we gently skip to the side and gather our feet:

> *right* 2 3 4 5 6, *left* 2 3 4 5 6, *right* 2 3 *left* 5 6, *right* 2 3 4 5 6.

In the fifth movement the circle dissolves and everyone moves freely through the room. The arms continue the arc movements out and up, freer, in flamenco or oriental style, one arm at a time or both, hands and fingers twist in and out from the wrist. The accents on the 1 in each measure are marked with Spanish or oriental hip movements. When two dancers or 'particles' come close to each other they aim a neat little nudge, hip against hip, making the two gently bounce away from each other. They end up close to another particle, a new little hip clash occurs—as the sound travels in waves.

Rehearse the five movements a few times before starting the whole dance. Then shape three circles:

> The air particles/sound waves are standing in three circles with their backs towards the loudspeaker or the congas drum in the centre, an inner circle, a middle, an outer.

The dance starts in canon. The circles do their first two movement frequencies in succession, two sets of each, right and left:

> The outer circle 'thumps' away from the sound source, doing the first movement, once with the right foot and once with the left. They spin around facing the centre and do the second movement on the spot, wrists and arms extending to the sides.
>
> Then the middle circle starts, moving like the outer circle: coming out from the centre, turning round facing the centre, doing the second movement on the spot, arms going to the sides. Then the inner circle starts off and does the first and second movements, like the first and second circles.
>
> By then the first circle has repeated the second movement twice, and the second circle once, waiting for the third circle to come out. Now everybody does the second movement together one time right and left: the middle circle does this joint set starting the second movement to the left, opposite to the first and third circles who start to the right.

When it is time for the third and fourth movement, the legs extend sideways, first on the spot and then with displacements, the outer and inner circles start off to the right, while the middle circle starts to the left, circling in the opposite direction to the outer and inner circles.

The fifth movement engaging arms and hips is done by everyone at the same time, during four phrases:

> The air particles move helter-skelter through the room, do their hip accents, bounce against each other, against walls, pillars, doors, loudspeakers, drums.

During the fourth and last phrase in the fifth movement with arms and hips, everyone gradually moves into one big ring around the sound source and does the fourth movement all together in the same direction, first with right arms and legs, then left, then right again, again and again:

> Meanwhile the circle widens. The movements become slower, the music tunes out and so do the sound waves. Left in the centre on the empty floor stands the sound source, silenced.

Meta-talk on Sound Recording and Playback

Discuss the transformation of sound waves into electronic and mechanical storage: analogue on vinyl records, reel tapes or cassette tapes; digital on cd-discs or other digital media. When played back, the stored medium is transformed back into sound waves, which, amplified, vibrate out from the loudspeakers like in the dance. Discuss the similar process where sound waves are received by the ear and the ear-drum, transformed via the hammer, incus and stirrup bones—the skeleton's smallest parts—into nerve impulses. These in turn are interpreted by the brain so we recognise the sound of instruments and voices, hand claps and stamps, rapids and thunderbolts.

Dancing to Instruments—Auditory, Visual, Tactile and Kinaesthetic Perception Interplaying

Hand out instruments to half the group. The musicians comprise four sections: percussionists on drums, tambourines, triangles; wind players on clay pigeons, whistles and flutes; some get rattles in different sizes; some get hard or soft plastic tubes. The other half of the group will dance.

Each of the sections of this orchestra works out a short music sequence, a rhythm they rehearse and memorise. The dancers work out a movement sequence to each sound sequence and rehearse them. The group makes as many sound loops and matching dance steps as they can manage. Repeat a few times in an order.

The musicians perform their music pieces a minute or so each, in any order. The dancers listen and dance their matching sequence. There is space to improvise, embroider and amplify the music as well as the dance. A new instrument breaks in, the movements adapt to the sound.

Eventually one orchestra section plays their instruments and the dancers dance. After a while another instrumental section joins in and the dancers find a way to dance the matching movements, alternating or simultaneously.

Meta-talk on Sensory Perception and Motor Activity

Why do we twitch and get ready to run when we hear a sudden loud noise nearby? The brainstem or reptilian brain takes over. The same happens when we want to pull away from a bad smell, a possible sign of an infection source. Or when we dodge a branch swaying towards us on a walking trail, or when we turn our face away as a reflex when someone is about to aim a punch. Movement is tied to shelter and survival. Even an amoeba scents minimal vibrations and moves to where their food is, bacteria and algae. Bats emit high frequent sound bouncing against winged insects and back, locating their food through sound.

Through our senses we communicate continuously with the world. The senses do their receptive work beyond our control—more than we can close our eyes, stop our ears and nose or close our mouth. Letting sounds, colours and movements penetrate our senses is an active–passive readiness, a capacity to be touched by the sensuous world before acting on the sensation. The work of the senses precedes thought. The 'sixth', kinaesthetic sense touches on the capacity to be moved by movement, which we practice when dancing, moving towards a richer interplay with the material, natural and social world.

Light

The Faceted Eye—Relay Dance

Music: driving upbeat, e.g. Cheb Aïssa: Chira France.

Send around different faceted glasses for everyone to look through. The image visible through the glass is repeated multiple times.

Everyone is sitting along the longer wall. Put on the music.

Someone gets up and dances freely to the music. After a while the music stops and the dancer freezes in a pose. The next one gets up and takes the same pose. The music continues and the new dancer moves in their own way and the first one hangs on.

The music pauses again, both freeze in the same pose. A third gets up and takes the pose. The music plays again, the third dancer takes the lead, the first and second imitate. And so it continues until everybody dances like the last one joining in.

The dance can also be done without music stops; then they all fill the floor one after the other.

At the end we see a bunch of dancer copies, all moving synchronically. This is the picture in the eye of a fly watching a human dancing. What is the fly's eye called? A *compound* eye or *faceted eye*. It consists of a few thousands of small hexagonal lenses collecting the light into one single optic cell, which the fly's brain puts together into one sharp image. How many lenses do we humans have? *Two*! One in each eye. Do we see everything twice? Yes, each eye sees an image which the brain puts together into one single picture. Thanks to having two eyes we perceive depth in what we see. Actually, we see everything up and down, but that too the brain fixes without us needing to think about reversing what we see.

Meta-talk on Light

What is light? *Waves*, like sound. The same kind of waves? Sound moves in *mechanical* waves of vibrating, bouncing particles of the medium which the sound travels through. Light moves in *electromagnetic* waves of *photons*, elementary particles who have an amount of movement energy.

Think about the sun's light illuminating our existence. Unrefracted light is white. When it hits surfaces and objects, parts of the light's spectrum is *reflected—shining back* or cast back—while other parts are absorbed—*sucked up* or taken in—which gives the subject or object its colour. Without light there are no colours. In the dark all cats are grey. Colour does not exist, only in our eyes. Colour presupposes light hitting matter, which reflects and/or absorbs the light. When we see a wall in white, that is because it is painted with a colour substance that reflects almost all light. When we see a sweater as black, it is thanks to the yarn being dyed with a substance that absorbs most of the light.

Hand out small prisms that refract the white light into its spectrum. So do water drops, individually or in moist air, conjuring a rainbow when the sun shines during or after rain.

Pick up the metal spiral used to demonstrate sound waves. Let someone hold one end and make wave movements again. The human eye cannot perceive all frequencies in the light's electromagnetic radiation, only the colours as in a rainbow. In what order do the colours appear? *ROYGBIV!* The *red* wavelength is the longest. Make big wave movements, and increase the frequency: after red comes *orange*, the wavelength decreases, then *yellow*, then *green*, then *blue*, then *indigo* and finally *violet*, the shortest wavelength in visible light. What comes before red, invisible to the human eye? *Infrared!* What comes after violet? *Ultraviolet!*

The Long Mirror Dance

Music: calm, without particular rhythm, e.g. Sigur Rós: Untitled #1 Vaka or Untitled #3 Samskeyti.

Everyone stands in two long lines across the whole length of the room, two by two in front of each other. During the whole dance everyone stays in their places in the two lines, their gazes kept straight ahead, fully concentrated on the dancer standing opposite. You always see a little in the corner of your eyes, but don't turn your head.

> The pairs do a mirror dance (see 'Mirror Exercises: Two by Two, Four by Four, Eight by Eight, All Together' in Chap. 3). They take turns leading. After a while both lead and follow equally much. After another while, without moving closer to each other, two pairs standing next to each other start dancing together four by four, synchronising their movements. A little later two quartets merge and dance univocally eight by eight.

The whole time everyone stays put in their places, without moving closer to each other, forward or sideways. Everyone's full *focus* is on the one right in front from the start. Encourage amenability. In this exercise taking in movement is more important than steering the course of events.

> Finally everyone is dancing synchronically, giving and taking, like one single big mirror dividing the room in the middle.

They find a common ending, letting the movements fade out and cease. It all takes the time it takes for the joint movement to unroll and come to an end, every move mirrored and reflected exactly opposite, in the 'mirror'.

Meta-talk on the Speed of Light

How fast do you get to see the movements of those standing opposite? *Instantaneously!* How fast do you catch sight of yourself coming in front of a mirror? *Quick as a wink!* At the speed of light. How fast does light travel? Almost 300,000 km per second in vacuum. When everyone moving in the mirror dance keeps a keen eye on the one in front, then how long does it take for a movement impulse to shift from one end of the room to the other, bouncing between eyes and mirror? Hardly any time at all.

The first person who tried to measure the speed of light was Galileo Galilei in the seventeenth century. In his experiment, two people were standing opposite each other, each with a lantern from which they alternatingly let the light shine and be blocked. They tried to synchronise their alternate movements, distancing themselves further and further away from each other in search of the limit for the light propagation time. The two people were fully focused on each other's movements, looking for the same synchronicity as we did in our mirror dance.

Kaleidoscope Choreographies

Pass around kaleidoscopes of different sizes and designs for everyone to look through. How do the patterns emerge? Small mirrors sit arranged in the other end of the tube, reflecting the coloured bits back and forth between the mirrors. As we rotate the tube the coloured bits tumble round and round and create new patterns again and again.

Divide the group into two halves. Each half create their kaleidoscope dance. Think from the perspective of a fly on the ceiling, looking down on the dance. Get inspiration from synchronised swimming.

Encourage free composition: Does everybody have to sit or stand in a polygon the whole time? Does everybody need to do the same movements the whole time? Is there space for small patterns in the outskirts? For small patterns within the big patterns? Encourage exploring details, using fingers, shoulders, faces. Hold on to a few fixed patterns and find ways to shift between them, like in a kaleidoscope.

Photosynthesis and Vesicular Breathing

Music: 2046 soundtrack/Main theme.

The final dance on the water, air and light themes is illustrative and dramatising. It refers to global heating as a consequence of the *Anthropocene*. Ask the pupils what they know about the significance of trees and green plants and their capability to take carbon dioxide and produce oxygen.

Discuss *photosynthesis* in green plants and the reverse photosynthesis, *vesicular breathing* or *cellular respiration* in human body cells. Write a simplified chemical reaction for the photosynthesis on the board, disregarding the exact number of molecules and the atomic balance:

$$6CO_2 + 6H_2O + light\ /\ the\ sun's\ energy \rightarrow C_6\left(H_2O\right)_6 + 6O_2$$

Trees and green plants take carbon dioxide from the air, and water from the earth, which they transfer, with the help by the sun's energy, into grape sugar and oxygen. Write a simplified formula for vesicular breathing:

$$C_6\left(H_2O\right) + 6O_2 \rightarrow 6CO_2 + 6H_2O + energy$$

The cells burn, meaning they *bind oxygen* to the grape sugar we eat, so we get energy, our cells grow, split and build new cells.

How does the oxygen in the air get into the body and find its way to the cells? We breathe in the air through the nose, mucous membranes and the windpipe into the lungs. Our lungs are made of air pipe trees, like motorways dividing into side roads, country roads, small roads, forest paths, ant paths and almost invisible air tubes into the pulmonary alveoli, the lungs' tiny small air sacs.

Split the group in two. Each finds a way to dance, one half the photosynthesis, the other half the vesicular breathing. Encourage all ideas springing coming up. For inspiration, remind them of the water molecule dance, the dance about the weather and the winds, and so on. Let each one find their role in this two-act drama—dance is best born out of dancing.

Put the music on.

> The sun goes up. Trees stand outspread in the landscape. Their branches and leaves sway in the wind. High up in the air the green leaves breathe in carbon dioxide molecules. Down in the ground the roots suck up water molecules. The sun shines through the tree crowns and triggers photosynthesis to start.
>
> An apple, full of fructose, falls from a twig. The leaves send oxygen molecules out in the air. The dancing trees transform into lungs breathing and hearts pumping. The O_2-molecules are drawn into the lungs. The lung breathes and sends the oxygen into the blood. The hearts pump the oxygen in the blood, round and round throughout the body. The oxygen molecules are absorbed by the cells who just helped themselves to the carbohydrates in the apple, burning them. The cells get new energy, swell and generate new cells, sending the carbon dioxide molecules back into the blood pumping towards the heart and lungs and into the exhaled air.
>
> The energy helps the body produce new cells, retain warmth and make sure we can move and dance altogether.
>
> The trees receive the carbon dioxide, suck up the water, make new oxygen with the energy of the light and a new apple of glucose.

Repeat the cycle with variations, refine, let the imagination flow.

Space

Satellite Dance

Music: calm, sustained, e.g. Pink Floyd: Shine on You Crazy Diamond, Pts. 1–7.

Create a relaxed atmosphere in a serene, concentrated, attentive warm-up.

> Everyone walks freely throughout the room, relaxed, in continuous contact with whole feet on the floor, straight backs and necks, arms hanging by their sides. When meeting someone, they round each other shoulder by shoulder and continue their trajectory.
>
> Without any other communication but kinaesthetic awareness, eyes are wide open, into the corners of the eyes, and 'eyes in the back of the head', everybody senses the group's joint movement. Lik one body, without anyone giving a sign or steering, the walking stalls and the group comes to a halt. Sense the stillness for a while, then start walking again. Repeat a few times with full concentration, in quiet interplay, the shift from movement to stillness becoming more and more attuned and instantaneous.
>
> Next, as everybody moves around in silence, one dancer slows down and stops. Right away, the others start walking in circles, everyone from wherever they happen to be, keeping the same distance from the one standing still. If a dancer's circular track leads into a wall, they

backtrack in their circle. The dancer standing takes in the sensation of being at the centre of the others' circling motion for a while, then walks on again, the circles dissolve with everyone walking at random in the space. Someone else stops and the others start circling around that dancer from wherever they are.

Repeat a couple of times so several dancers get to experience being in the centre. Everyone is observant, no one knows in advance who will come to a stop. If you're about to stop and notice another dancer has become immobile, start walking in your circle around the one who stopped first.

Next, two dancers stop, a bit away from each other. The others begin their circular tracks, either around the one standing nearest, around both, or alternating between the two, all depending how close they get in their trajectories.

In the next variation only one dancer stops, and starts spinning round their own axis. The others instantly move into their circles. After a while the one in the centre goes back to walking around freely, and the circles dissolve. Another dancer stops, rotates, new circular tracks emerge.

In the last variation, one dancer stops, then starts revolving around their axis and moves in their own circle. Everyone else enters their circular tracks and trails the one moving in the centre of all the motion.

Meta-talk on Satellites, Planets and the Solar System

What is the word for a body circling around another body? A *satellite*. What is it that keeps one body in a track around another one? Welcome different terms and explanations to do with satellites, both bodies that have existed in space before humans and those sent up into space by humans: *parabola*, *gps*, *gravitation*, *mass*, *rotation*, *orbit*, *celestial body*, *star*, *galaxy*, *planet*, *moon*, *meteor*, *comet*, *astronaut*, *cosmonaut*, *satellite states*, *space station*, *manned orbiting laboratory*, *space shuttle*, *spacewalk*, *space probe*, *telescope*, and so on.

How many planets orbit around the sun in our solar system? Eight. Nine? Discuss *Pluto*, which no longer qualifies as a planet, but whom we'll welcome to the dance in a while.

The word planet comes from Greek *planetes*: *wanderer*. What is that tremendous force of attraction that the sun uses to hold together the whole solar system, that colourful group of planets? The sun's mass. It makes up 99 percent of the solar system's total mass. The rest, all the planets and their moons, the asteroid belt and the dwarf planets, together make up the remaining 1 percent.

The Planets' Dance

Props: a yellow ball, nine dresses/skirts/draped pieces of cloth in different colours, four signs with the dates 21 December, 21 March, 21 June, 21 September, pictures of the planets, the sun, galaxies.

Presentation of the Planets

A yellow ball lies on the floor, leaving enough floor space to line up all eight planets plus Pluto in a row starting from the sun.

Which planet is the closest to the sun? *Mercury*. Who does Mercury? Someone puts on the first, greyish garment and goes and stands near the sun. From which language does the name come? From Latin. What kind of name is it? In Roman mythology *Mercurius* is the *messenger of the gods*. The Romans took over the Greeks' mythology, where this swift-footed messenger with winged sandals was named *Hermes*. What does the name hint about the planet's speed? Mercury is quick, he darts away over the vault of heaven with the latest news and gossip about the gods' plots and intrigues troubling each other, half gods and humans. Discuss the planet's colour, greyish; its area, like the moon's with craters; what minerals it has on the surface; if it has an atmosphere; and how cold or hot it is on Mercury. Does it have a moon? No.

What is the next planet from the sun? *Venus*. Who does Venus? Someone puts on a cream-coloured dress and stands next to Mercury. What is her name in Greek, she is a she, is she not? *Aphrodite*, the *goddess of love*. This is the earth's twin planet, almost the same size, with a carbon dioxide-saturated atmosphere, clouds of sulphur dioxide and thus a devastating greenhouse effect. It is hot on Venus, 500°, like a pizza oven. You don't want to land there in search for love and suffocate in her sulphurous acid veils. Why is Venus called the *morning* or *evening star* in many languages? Mercury is also called this. How come? When do the two planets between us and the sun become visible from Earth? When the sun has gone down or before it rises. Does Venus have moons? No.

Go on to the next planet. *Earth*, *Mother Earth*, *Tellus* in Roman mythology. Who wants to do Earth? The blue planet, why is it called that? The Earth has an atmosphere and lots of water, as we well know. This, along with a fair distance to the sun has catered for life to arise on Earth. How many moons does our planet have? One. What is moon in Latin, in Spanish? *Luna*.

Which planet comes next, after Earth? *Mars*, the red planet. In Greek it is *Ares*, the *god of war*. The dancer ready to do *Mars* puts on a reddish dress and goes to stand next in line. Why does it look red? It is covered with iron oxide dust and its whole atmosphere is *rusty*. How many moons does it have? Two, Phobos and Deimos, phobia and demon, Fear and Terror of war, twin sons of Ares—and Aphrodite.

What is this group of four planets the closest to the sun called? The *terrestrial planets*, the *telluric* or *rocky planets*. Unlike the *gaseous*, *giant planets* coming up next.

Between the two groups of planets there is a ring of stony bodies, orbiting and held there between the sun's and Jupiter's gravitation: *the asteroid belt*.

Which one is the first gaseous giant? *Jupiter*. In Greek? *Zeus*, the *chief god*. Who does Jupiter? The biggest planet, with a mass more than twice and a half the mass of all the other planets together, yet less than one thousandth of the sun's mass. How many moons does Jupiter have? Over ninety, and new ones are still being discovered. *Galileo* observed the largest four with his telescope, built in 1610: *Io*, *Europa*, *Callisto* and *Ganymede*, all four bigger than Mercury. In Greek mythology *Europa* was a Phoenician princess whom Zeus abducted, and *Io* was a Greek princess with whom the same Zeus fell in love. And that's how the plots and treachery, the abduction of daughters and revenge, the conquering of land and waging of war between people and gods in all myths and times got going.

Next planet? *Saturn*. *Cronus* in Ancient Greece, the *god of sowing and harvest*, father of Jupiter. What is so special about Saturn? It has rings of ice and comet dust. The next dancer puts on a fabric with stripes of ribbons and joins the line. Jupiter, Uranus and Neptune also have thin rings of ice.

Uranus and *Neptune*, the ice giants, are the last planets in line. In Greek *Uranos* and *Poseidon*, the *god of the sky* and the *god of the sea*. Both shimmer in blue and turquoise. Besides hydrogen and helium they contain a great deal of *methane* in their masses and in their atmospheres, which absorbs red and reflects blue in the light's spectrum. Who wants to do these planets? Two dancers take on blue garments and join the line of planets.

And so last in our line is *Pluto*, now moved to the dwarf planets' league. But Pluto is welcome to join our dance. Many children know and have feelings about Pluto. Do we have the heart to let the dwarf planet wander around far away in cold space all by itself? It was that darkness and coldness that made an eleven-year-old girl in Oxford, England, propose a name after the planet had been discovered in 1930: *Pluto*, the *god of the underworld*, in Greek mythology *Hades*, *kingdom of the dead*. Does little Pluto have moons? Yes, five, for what is known.

The Planets' Dance Around the Sun

All planets dressed or draped in their fabrics are standing in a row starting from the sun. The others sit down and clap the beat to the planets' dance around the sun. Out in space the planets are never lined up this neatly. They move round in their own elliptic orbits in different planes around the sun, the further away the wider orbits. For visualisation, we can imagine concentric circular tracks in one plane. The further the planets are out in space, the longer it takes them to complete one orbit.

Lay out four sheets of paper with dates 3 months apart. First 21 December, *winter solstice*, at the far end of the line from the sun, a bit beyond Pluto. That day of the year Earth's axis leans the furthest away from the sun in the northern hemisphere, making the day there the shortest, and the other way around in the southern hemisphere. We move on a quarter of a year's orbit. Put the date 21 March, *spring equinox*, on an axis through the sun, at a right angle to the line with the planets and the previous date. This is the point of time where day and night are equally long in all time zones around the globe. We move on another quarter of a turn, to 21 June, *summer solstice*. Place the sheet at the far end of the line from 21 December, through Pluto and all the planets and the sun. At this time of year on Earth the conditions are opposite to the winter solstice: the earth's axis leans the most towards the sun in the northern hemisphere, making the days the longest north and the shortest south of the equator. Finally lay out the sheet 21 September, *autumn equinox*, another quarter of a year's turn ahead, at the other end of the line from *spring equinox* through the sun.

Now we have four orientation marks to explore the planets' orbits around the sun, which are tied to the Earth's movement over one whole Earth year, twelve Earth months.

The Earth steps out of the planets' row and takes twelve strides in a circle around the sun, one stride for every month. The starting position is in line with 21 December, three steps to come in line with the spring equinox, another three months to the summer solstice, three steps on to the autumn equinox, and another three months to 21 December one year later.

How many times does the Earth spin around its own axis during its year's orbit? 365 times, roughly 30 times a month—which we won't ask from Earth in this dance, unless they want to try.

Someone sitting beats the rhythm on a drum, the others clap along. Count: *ten … eleven … twelve …* and count along loud from *one* to *twelve* while the Earth takes its twelve monthly strides around.

So what about Mercury? Does it move faster or slower than Earth? *Faster*. Since it is closer to the sun, its orbit around is shorter. Does that entail more or fewer turns around the sun during one terrestrial turn? *More*. How many laps does Mercury complete? *Four*. Then how long is a Mercury year? Three months on Earth.

Those sitting count *ten … eleven … twelve …* and *one …*!

Earth and Mercury take twelve steps around the sun. Mercury comes around in three strides, completing a set of four orbits, while the Earth finishes one orbit.

How many laps does Venus complete in one Earth year? Roughly one and a half laps around the sun. In our planets' dance Venus ends up in the position for *the summer solstice*. How long is one Venus year then? About 8 months. Next in line is Earth, which we have the hang of.

What about Mars? Does Mars move faster in its orbit around the sun one turn, or does it take longer time? *Longer*. In one Earth year Mars only reaches about half a turn around the sun. How far does Mars get after twelve monthly strides? To the *summer solstice* position. So how long is one Mars year? *Two* Earth years.

Jupiter? It takes more and more time to complete one orbit around the sun for the planets next in line from the sun. Jupiter needs twelve times twelve steps to come around once, that is 12 years. One Jupiter year is twelve Earth years. In our dance Jupiter reaches a position around 21 January, the same position Earth reaches in one step.

After one Earth year Saturn has reached a position for 6 January, Twelfth Night, meaning one Saturn year is 30 Earth years. Uranus gets to the position for 25 December, Christmas Day, and needs 84 Earth years to revolve once around the sun. Neptune only reaches the Christmas Eve position, spinning around the sun in 165

Earth years. Pluto hardly leaves its spot, getting one and a half day on from the winter solstice position, the dwarf planet's year lasting almost 250 Earth years.

Now all planets know their part in the solar system choreography. Those sitting count the beat on the drum or clap their hands: *ten … eleven … twelve …* and *one … two …* on to *twelve*.

The planets take twelve steps, Mercury taking the largest strides, the gas planets taking their smaller and smaller steps and Pluto hardly getting anywhere on twelve tiny steps. In order to clearly mark their progress, they all lift their knees high for every step, one Earth month at the time.

The planets do their dance around the sun one more time. Someone hops in with a hand drum to do the Earth's moon. How long time does it take for the moon to go one turn around the Earth? Twenty-eight days, almost a month: about one turn for every step the Earth takes. This can be marked with the hand drum circling above the head of the dancer doing the Earth, or that dancer moving a ball around their hip for every step.

What if the planets while moving forward also rotate on their axes, like we did in the warm-up. In which direction does the Earth rotate? Think about how the sun moves through the sky from sunrise till sunset … Then the rotation must be … anti-clockwise! Do all the planets spin like that? Do some go counter to the others? Venus, mad with love. And Uranus, who's rotation axis almost lies parallel to its orbit, so the planet almost rolls in its far away orbit around the sun.

And what about the sun, does it stand still in space? Or does it also move around a celestial body? It would have to be something with a tremendous mass and a staggering gravitation. Show pictures of *galaxes*, like the *spiral galaxy The Milky Way* and our next neighbour *Andromeda*, and billions of other star systems with billions of stars, each star with its own planetary system.

Our sun with our planets moves on the fringes of the Milky Way, which itself revolves like a giant porridge plate around a big *black hole* in the centre, *Sagittarius A**. It has an awfully big mass and attractive force, not bigger than our solar system but a mass four million times bigger than our sun. How long time does it take for the sun to fulfil one orbit around the black hole? 240 million years, barely a quarter of a billion years.

How many times has the sun orbited around the centre of the Milky Way since it came into being? Scarcely twenty. How many turns does it have left before it burns up? About the same number. Turns left …? Burns up …? In the next dance we go into the sun's fate as a star and the mind-blowing perspectives on our humble existence in space and time.

Continue discussing, depending on the group's previous knowledge or possible ongoing thematic work: space; the size of the sun, the Earth, the moon and the other planets; the distance to the sun in *light minutes*, *light hours* and *light years*; real and apparent *retrograde* movements; and so on.

Fusion Dance in the Inner Sun

Music: driving rock, e.g. Survivor: Eye of the Tiger.

Everyone is standing in a circle. Let's imagine we form the sun's *corona*, its visible atmosphere. Everyone holds their arms up in a right angle, like an H. What was H in chemistry language? *Hydrogen.* The sun consists of almost three quarters of hydrogen and almost one quarter *helium*: *He*. The rest is oxygen, carbon, iron, neon, nitrogen, silicon, magnesium and phosphor.

This time we think of ourselves as *hydrogen atom nuclei*, not whole H-atoms as the two in a water molecule. An atomic nucleus sticks tightly together by a big binding energy. When somehow an atomic nucleus is split, a huge amount of energy is released. What is splitting or division called? *Fission.* As in a nuclear power plant, where the heavy nuclei of uranium atoms are split, releasing strong nuclear radiation.

It is equally difficult to *smash together* two atomic nuclei. What is that amalgamation called? *Fusion.* The difference is that fusion does not leave any radioactive residual waste products behind for thousands of years, only pure energy. That is why technologists work hard hoping to 1 day build such power plants.

In the centre of the sun the temperature is at least 15 million °C, and there is tremendous pressure. This creates conditions for fusion: the merging of four hydrogen atomic nuclei into one helium atomic nucleus. The mass of the helium nucleus composed is smaller than that of the four hydrogen nuclei together. What is the relation between mass and energy in the universe, according to the *principle of relativity* or *reversibility*? Mass is a form of energy that can transform into other forms of energy.

Which formula expresses this energy generation? *Einstein's equation:* $E = mc^2$, where m is the mass reduction and c the speed of light in vacuum: 300,000 km per second. When four H-atomic nuclei fuse into a He-atomic nucleus the loss of mass is roughly 0.7 percent. Not as big a deal as one might think. The fact is that every single second 600 million tons of hydrogen is transformed into helium ashes, which means that roughly four million tons of the sun's mass is transformed into pure energy in the inner sun's hearth, every second. Furthermore, it takes one million years for the energy to radiate out from the sun's core to its surface and on into space, as *sun winds*, towards Mercury, Venus, the Earth and all the way to Pluto and further out into the cosmos. With the sun winds, one million tons of sun mass burn up per second.

Put the music on.

Everyone is standing in the circle like H-atomic nuclei, shaking and rocking to the music. Call 'fusion!' Four hydrogen atomic nuclei rush to the middle, lump together in a big hug, and sweep out from the centre, through the solar body and the corona, as light and heatwaves, out into space. Call 'fusion!' again. Another four H-nuclei rush forward, clod into a He-nucleus, and radiate out. Continue until there is no-one left in the circle, or less than four dancers.

The dance shows how the sun constantly burns up mass—five million tons per second transforming into solar energy. For how long has this been going on? Since the sun was created, some 4.5 billion years ago. What does this deficiency of mass mean for gravitation? The sun's cohesion and gravitation diminish. How long can this go on before the sun has burnt up all its hydrogen? As many more years. Then the helium fuses into carbon and the sun starts to swell.

When gravitation decreases the sun starts having problems keeping itself and the solar system together. The sun swells onto a *red giant*. Eventually it may swallow Mercury and Venus. However, due to the weakened gravity, the planets' orbits have widened, and Earth has wandered further away. When at last the helium is used up altogether, the sun collapses into a *white dwarf* and goes on crumpling into a *black dwarf*, a dense, hard and cold celestial body.

Such is the life cycle of a star like our sun. But as said, it has another 4.5 billion years to go. Meanwhile, we get to live our lives, play and dance through school, learn lots and dance for many years with all kinds of people we meet and become fond of. We can take care of the planet and hand it over to the next generation, and that one to the next, and so on. We get to dance and ponder over many issues about what life on Earth is about. Maybe one of you will travel to the moon, or to Mars. Maybe we'll encounter new acquaintances out there in space and widen our view even more.

Thinking about how long this has been going on, let's backtrack into the history of space. Back to where it all began, to that *big bang*.

Matter and Antimatter—Dancing the Nobel Prize for Physics 2008

Music: driving heavy rock, e.g. Lynyrd Skynyrd: Sweet Home Alabama.

Everyone stands in a circle again. Each dancer is allocated a letter: M, A, M, A … Make sure the last one is an M so there are more Ms. Step in as an M yourself.

> The Ms make themselves big and voluminous, like hulks, elbows sticking out, blowing up their cheeks. They rumble and clump around, taking up a lot of space. They are self-important, they are solid, they are tangible: they *are*. The Ms are *matter*.
>
> The As are vague, formless, sly and slimy types. They move evasively, like shadows, they are sort of negative pictures of the Ms: as if they *were not*. The As are *antimatter*.

Turn on the music.

> Everyone starts rock-'n'-rolling. The Ms in their fashion, the As in their way. Allow time for both kinds to establish themselves and mingle around. Eventually one of each sort is drawn to the other. An encounter occurs. They hug, but it is formally a kiss of death: they annihilate each other and sink down on the floor, obliterated, dissolved into each other and forever gone. Another two of each kind meet, hug, vanish. Until only Ms are left. They make the victory sign.

Meta-talk on Matter and Antimatter

What we danced is precisely what evolved just after *the big bang*. There was a tremendous amount of matter and antimatter. As the universe continued to expand the two kinds took out each other. They were *annihilated*. The conditions were, however, *asymmetrical*: there was a surplus of matter, why there was something left at all. As time went on, chemical elements were formed out of the matter. Then these merged into molecules, which on Earth eventually formed living organisms, first one-celled, later multicellular. Life arose in the oceans, then on land: fish, birds, mammals. Humans cultivated the soil, made tools, machines, satellites, learned and learn about nature and space, love, dance … and continue asking questions about how everything hangs together and where we are in all this.

Those who worked this out were two Japanese and one American scientists of Japanese origin, who were awarded the 2008 Nobel Prize for physics for their joint groundbreaking discovery. Nonetheless, how everything evolved in the beginning of time is still in many respects a mystery.

From the Big Bang to Aniara

Music: cosmic theme, e.g. Pink Floyd: Shine on You Crazy Diamond, Pts. 1–7.

The galaxes in the universe move further and further away from each other: the universe expands. If we think back in time, everything must have once been closer together. The *big bang* theory is generally established. Initially the entire observable universe was concentrated in a little pea. Then everything expanded rapidly. Space, time and matter arose, and all matter constituting the atoms, stars, planets and galaxes expanded to the current size of the universe. This expansion continues today, and it does so faster and faster.

Everyone sits spread out on the floor. Play music in the background. Read a text, a poem, a monologue or some lyrics on an existential theme. Below is the preface to the epic poem ı from 1956 by Harry Martinson.

Everybody listens to the background music and the recited text. Then they rise slowly and, millimetre by millimetre, move closer towards each other in the middle of the room, cramming together as if inside a tiny little pea.

Aniara—a review of humankind in time and space—is the story of the spacecraft Aniara, evacuating eight thousand people to Mars, after an environmental and nuclear weapon disaster threatening all life on Earth.

An emergency veer in the asteroid belt forces the shuttle off course, further and further away, until no return is any longer possible.

Aniara is about everything we do not personally master but which we nonetheless are a part of and belong together with. In whatever way we live our lives, it evolves within frames that are in some way or the other inexorably determined. One of these frames is the biological.

> We are born, we mature, grow old and die. Already this frame encompasses all joy and fear. Furthermore, we live within the frames created by humankind in interplay with nature: the social, the political, the religious and the scientific. Our entire lives are a series of attempts, within these frameworks, to explain the world to ourselves, to the limits of enigma or dread, or to shut ourselves off from it and protect ourselves by means of symbols of introversion and metamorphoses of instinct.
>
> To say that we need to make the best of this global situation is an old truth. Yet our world, and our image of it, has in less than one century—in a few decades—developed in ways even the seemingly obvious is shaken up. What previously appeared as manageable and evident has swollen and now approaches the enigmatic and the tremendous. With all our sheltering frames around us we have been launched towards infinity.

Martinson calls the text of his epic poem *Aniara* a product of imagination written by time. A poem about our common Earth and our hopes for the world, our sorrows and our disappointments, but also about our efforts to create respite facing an implacable course of events, with the help of imagination.

The text brings us back to the existential and ethical dimensions of our search for knowledge and meaning. We do not know everything about all that surrounds us, only that it is something bigger, something that has been in existence longer than humankind and that to all appearances will survive us. We need to attune and interact with these tremendous forces and resources and take care of our world during our short time in it. Written 70 years ago, the text offers an uncanny topicality regarding our dialogue with the world today, the natural, material and social world.

Martinson contributes to our efforts to understand existence through writings that 'catch the dewdrop and reflect the cosmos', as stated in his motivations to the Nobel Prize in Literature in 1974. Literature, art and dance can give us perspective and guidance, make us stop and sharpen our senses. They can turn our attention to what there is to see and experience, interpret and better understand, give meaning, or just let it touch and astonish us—give us a glimpse of what remains unknowable. With our imagination and our senses we can grasp our place and possible contribution to that big wonderful and disquieting universe—and what it demands of us humans.

The Last Dance

The last dance is an improvisation in slow motion on the themes of the *big bang* and the consequential evolution of the solar system, including life on Earth. The initial movements are minimal and slow. As the space for manoeuvre increases so does the scope of the movements.

Some dance matter-antimatter. Some expand into the first elements taking shape during the first tens of millions of years. First the lightest—hydrogen, heavy hydrogen and helium. Then nitrogen, silicon, iron, methane, carbon. The elements merge into chemical compounds like carbon dioxide, salt, nitrates and water.

Bacteria are formed. Symbioses of two bacteria occur a couple of billions of years ago and develop into mitochondria, which all animal cells contain, and into chloroplasts with chlorophyll, which can transform light into chemical energy.

Unicellular, shelled fauna organisms are formed 540 million years ago; trilobites, from 520 million years ago; and the first fishes, 480 million years ago. The first land plants develop 440 million years ago; the first insects, 415 million years ago; and the first reptiles, 360 million years ago.

Trilobites and multiple sea fish die out 250 to 300 million years ago. A big meteorite may have crashed onto Earth. The Permian-Triassic extinction disaster, also called the *Great Dying*, occurs 250 million years ago, killing 70 percent of terrestrial vertebrate species and 80 percent of marine species, assumed to be due to volcanic activity lasting some 50,000 years. The first dinosaurs enter the scene 200 million years ago; the first birds and blooming plants, 145 million years ago; the first mammals, 65 million years ago; and the first human beings, 2.6 million years ago.

Generation upon generation of humans live among trees, plants, water, and not least oxygen, provided through photosynthesis for fish, birds and humans to breathe. Humans invent the wheel, steam engines and airplanes, build rockets and space crafts, solar panels and wind power plants, computers with algorithms and artificial intelligence. Everything keeps expanding, faster and faster.

Some dancers disappear out of the room, waving goodbye.

Discuss the origin of life, the origin of species, humankind's place in universe. Discuss science and faith, art and education, dance and teaching.

Chapter 7
Letting Dance Teach: A Call for Aesthetic Teaching

P. Moerman, *A K-12 Dance Method to Teach Creative Dance, English, Math and Science*, https://doi.org/10.1007/978-3-031-93156 7_7

This book on teaching dance inquires into what dance can teach us, also about education. Its theoretical line of thought departs from Gert Biesta's ideas on letting the arts teach: to show pathways to dialogue with the world. Trying to enter such dialogue is what education is about.

I tie the theme of dialogue between the self and the world to Biesta's plea for an education that takes on its existential remit: to provide space for the young person to break into the world as subject. I weave these threads of thought together with John Dewey's ideas on aesthetic experience and Susan Stinson's stands on dance education, kinaesthetic experience and dancing as becoming, informed by the daily practice of dance teaching.

First and foremost, this book advocates providing space in schools for young people to dance together, to draw on their power of initiative and imagination in movement, to create and embrace the unforeseeable and the unknown—the 'not yet' and the 'not yet known'—in dance, and to be taught by what is beyond one's grasp—which is the point of education.

More broadly, the book forwards dance teaching as a model of *aesthetic teaching* in a plea for an *aesthetic turn* in education: to take the arts seriously as sources and modes of knowing, insight and meaning-making. Aesthetic teaching calls the capacity to be affected by the senses into play, a readiness to be taught by and to act on what comes to us through the senses.

Teaching is about pointing, says Biesta. The teacher draws the pupil's attention to something out in the world worthwhile paying attention to. The pointing gesture is an invitation to be initiated into the realm of meaning-making. It is for the student, as subject, to accept the invitation, to be willing to be turned towards the world and to be appealed by what is pointed at out there. It is for the teacher to make the invitation tempting. In order to elicit a true and free response from the student, Biesta argues, the teacherly work must be aesthetic. Herbart, who coined the interaction between teacher, student and the matter pointed at as *the didactic triangle*, claimed that the teacher's mission—indeed the *art* of teaching—is to allow the world to reveal itself aesthetically.

Years of teaching creative dance and dance integrated in subject teaching have given numerous testimonies to the tremendous power and essence of this work of the senses. When young people dance together, engaging in kinaesthetics, they appear before each other, meet each other anew and build delicate ethical and existential relationships, imbued by joy of inquiry, sensitivity, proximity, trust, participation and communication—'wordless dialogues of giving and taking' as they word it.

In creative dance, with no other end or meaning but exploring dance and movement's own significance, the work evolves by surrendering to kinaesthetics. Also in dancing language, math and science, the didactic triangle is brought into being primarily by dancing and attending to the kinaesthetic. The (kin)aesthetic work precedes the cognitive, which in turn is given ample space while discussing the subject matter the dances reveal. Not only does this order lower thresholds to subject matter many find difficult. This aesthetic order is how we encounter the world in the first place, how we *receive* the world, before interpreting it and giving it meaning.

The key capacity at stake is captured in Wolff-Michael Roth's notion *passibility*: the capacity to be permeated by what comes to us through the senses. It is an active sensuous responsiveness to what the world offers as 'gifts'. Biesta indeed speaks of 'gifts' of teaching as opposed to controlled instruction aimed at the 'learning' of fixed curriculum content and a view on knowledge acquisition objectifying the world as a resource stockpile.

Instead, Biesta calls for an education 'beyond learning', providing space for the existential, freeing teaching from result-oriented goals and bringing the 'I' of the student into play: the subject's freedom. Any tendency to control learning processes and measurable outcomes jeopardises this freedom, by sticking to the linear and the predetermined and keeping away from the unpredicted and the unexpected.

Not everything can be predetermined and foreseen, not in life, not in teaching, not with respect to what may appear as unknowable. This is what artists bother themselves with. The idea of knowledge and insight as beforehand 'given' is objectifying of the student and their interaction with the world: it robs human experience from what *possibly* may emerge as surprising 'gifts'.

The opposite order to encounter and get to know the world is to let oneself 'be taught', to let oneself 'be summoned' to experience what there is 'out there', larger than what I can grasp and contain, ready to give itself to me: a subjectification of the one experiencing and of the experience offering itself. This order presupposes receptiveness and risk-taking, to see without foreseeing, a readiness to subject to what comes my way, including things I did not ask for or did not even know to ask for.

This order turns schools and education into scenes of revelation, says Biesta, realms beyond one's horizon of understanding, with a potential to strike with a sudden new insight. More than we may wish to learn about the world, the world is trying to give itself to us, a givenness preceding our attempts to grab hold of the world.

Dance prompts us to let go of the familiar and step into the unknown and the unforeseeable—an educational opportunity. Dancing and handling resistance in movement, of bodily limitations, limits of time, space and force, is an act of open-ended, inquiring dialogue, filled with existential encounters. In dance, *passibility* translates into the capacity to be moved by movement, a union of passive-active responsiveness permeating every instant in all dance activity dealt with in this book. To teach dance is to offer space for kinaesthetic experiencing, a pre-cognitive, pre-emotional, sensuous way of connecting to the world. Stinson views kinaesthetic experiencing as the sensuous internalisation of movement, which allows us to feel others' dance and our response to it.

Dance is transforming dialogue stripped from any technical utility, down to the 'naked' human body-subject interacting with the world, vulnerably and powerfully, aesthetically and ethically. Dance mediates the eternal value and meaning of mere movement—as eternal as the infant trying to roll over or the waves rolling back and forth on sandy beaches and rock formations around the planet.

There is nothing mystical about this. As Merce Cunningham states, dance is essentially about every single instance as it occurs, and dance's life and force lie precisely in this particularity. Dance is just as explicit and transient as breathing, as are the breaths of winds brushing over cliffs and homes and rocking the waves—or the waves rocking the winds to rest.

Dance education, from early childhood, is about handling physical and abstract elements like body, space, time, force and relationships, a hands-on work, says Stinson, through which children see that these things exist in dance, and in the world around. They can dance slowly like clouds, they can move swiftly like winds. Dancing and interplaying with the world, they encounter themselves. Dance 'teaches' us to move through and along with the world of things and creatures.

So teach and let young people dance and connect to each other and let the natural and social world in which we all exist reveal itself to our senses with all its gifts to cherish and care for. Let the children and students, with their imagination and creative force and on equal terms, enter the relational world of dance—a magic world, as Stinson's young dancers call it, of becoming.

That is what dance can teach us, if we are willing to let dance teach.

References

Arendt, H. (1954). *Between past and future: Eight exercises in political thought*. Penguin Books.

Biesta, G. (2006). *Beyond learning: Democratic education for a human future*. Paradigm Publishers.

Biesta, G. (2012). Giving teaching back to education: Responding to the disappearance of the teacher. *Phenomenology & Practice, 6*, 35–49.

Biesta, G. (2013). *The beautiful risk of education*. Paradigm.

Biesta, G. (2017). *Letting art teach. Art education 'after' Joseph Beuys*. Arnhem: ArtEZ Press.

Biesta, G. (2021). *World-centred education. A view for the present*.

Biesta, G. (2024). *Lad kunst undervise 1&2*. Lectures, VIA University College, Aarhus. [*Letting art teach*, Danish translation release lectures]

Bishop, A. J. (1988). *Mathematical enculturation: A cultural perspective on mathematics education*. Kluwer.

Comenius, J. A. (1991/1658). *Orbis Sensualium Pictus*. V+K Publishing.

Cunningham, M. (1952). Space, time, and dance. In *Trans/Formations 1*. Wittenborn & Co.

Cunningham, M. (1955). The impermanent art. In *7 Arts No. 3* (pp. 69–77).

Dewey, J. (1916). *Democracy and education*. Macmillan.

Dewey, J. (1934). *Art as experience*. Penguin/Perigee Books.

Dewey, J. (1938). The philosophy of the arts. In J. Dewey (Ed.), *(2008/1988). The later works, 1938–1939* (Vol. 13, pp. 357–368). Southern Illinois University Press.

Gelman, R., & Gallistel, C. R. (2009). *The child's understanding of number*. Harvard University Press.

Greene, M. (1986). The spaces of aesthetic education. *Journal of Aesthetic Education, 20*(4), 56–62.

Greene, M. (1995). *Releasing the imagination: Essays on education, the arts, and social change*. Jossey-Bass Publ.

Greene, M. (2001). *Variations on a blue guitar: The Lincoln Center Institute lectures on aesthetic education*. Teachers College Press.

Greene, M. (2007). *Aesthetics as research*. The Maxine Greene Center for Aesthetic Education and Social Imagination.

Hagtvet, B. E. (2006). *Språkstimulering. D. 2, Aktiviteter och åtgärder i förskoleåldern*. Trans. Ingrid Häggström. Natur & Kultur. [*Literacy stimulation. D. 2, Activities and measure in preschool age.*].

Herbart, J. (2017/1902). *Science of education: Its general principles deduced from its aim and the aesthetic revelation of the world*. Trieste Publishing.

Joyce, M. (1994). *First steps in teaching creative dance to children*. Mayfield.

Levinas, E. (1969). *Totality and infinity: An essay on exteriority*. Duquesne University Press.

P. Moerman, *A K-12 Dance Method to Teach Creative Dance, English, Math and Science*, https://doi.org/10.1007/978-3-031-93156-7

Levinas, E. (1987). *Time and other and additional essays*. Duquesne University Press.

Levinas, E. (2003). *Humanism of the other*. University of Illinois Press.

Manchev, B. (2011). Vermögen, Ausbeutung und Widerstand der Subject-Körper. Für eine transversale Veränderung. In R. N. Lorey & G. Raunig (Eds.), *Inventionen* (pp. 197–213). Diaphanes.

Manning, E. (2009). *Relationscapes: Movement, art, philosophy*. MIT Press.

Manning, E., & Massumi, B. (2014). Just like that: William Forsythe: Between movement and language. In E. Manning & B. Massumi (Eds.), *Thought in the act: Passages in the ecology of experience* (pp. 31–58). University of Minnesota Press.

Marion, J.-L. (2017). *The rigor of things. Conversation with Dan Arbib*. Fordham University Press.

Martinson, H. (1956). *Aniara: A revy om the human people in time and space*. Bonniers.

Marton, F., & Tsui, A. B. M. (Eds.). (2004). *Classroom discourse and the space of learning*. Lawrence Erlbaum Associates.

Moerman, P. (2016a). *Dancing math. Teaching and learning in the intersection of aesthetic and mathematical literacy. The Bridges 2016 Finland proceedings on mathematics, music, art, architecture, education, culture*. University of Jyväskylä/Phoenix: Tessellations Publ.

Moerman, P. (2016b). Dancing and learning natural sciences. Aesthetics, embodied knowledge and classroom interaction. *The International Journal of Science, Mathematics and Technology Learning, 23*(3), 15–32.

Roth, W.-M. (2011). *Passibility: At the limits of the constructivist metaphor*. Springer.

Stinson, S. W. (1988). *Dance for young children: Finding magic in movement*. AAHPERD.

Stinson, S. W. (1990). Dance education in early childhood. *Design for Arts in Education, 91*(6), 34–41.

Stinson, S. W. (2004). My body/myself: Lessons from dance education. In L. Bresler (Ed.), *Knowing bodies, moving minds* (pp. 153–167). Klüwer.

Stinson, S. W. (2005). Why are we doing this? *Journal of Dance Education, 5*(3), 82–89.

Stinson, S. W. (2016). *Embodied curriculum theory and research in arts education: A dance scholar's search for meaning*. Springer.

Strindberg, A. (2008/1912/1902). *A dream play*. Trans. Edwin Björkman. Charles Scribner's Sons/ US Archive. https://ia601303.us.archive.org/19/items/playsbyaugust str00stri/playsbyauguststr00stri.pdf

Index

P. Moerman, *A K-12 Dance Method to Teach Creative Dance, English, Math and Science*, https://doi.org/10.1007/978-3-031-93156-7

MIX
Papier aus verantwortungsvollen Quellen
Paper from responsible sources
FSC® C105338
FSC
www.fsc.org

If you have any concerns about our products,
you can contact us on
ProductSafety@springernature.com

In case Publisher is established outside the EU,
the EU authorized representative is:
Springer Nature Customer Service Center GmbH
Europaplatz 3, 69115 Heidelberg, Germany

Printed by Libri Plureos GmbH
in Hamburg, Germany